T0133714

Multi-Carrier Communication Systems with Examples in MATLAB®

A New Perspective

OTHER COMMUNICATIONS BOOKS FROM AUERBACH

Analytical Evaluation of Nonlinear Distortion Effects on Multicarrier Signals
Theresa Araújo
ISBN 978-1-4822-1594-6

Architecting Software Intensive Systems: A Practitioners Guide
Anthony J. Lattanze
ISBN 978-1-4200-4569-7

Cognitive Radio Networks: Efficient Resource Allocation in Cooperative Sensing, Cellular Communications, High-Speed Vehicles, and Smart Grid
Tao Jiang, Zhiqiang Wang, and Yang Cao
ISBN 978-1-4987-2113-4

Complex Networks: An Algorithmic Perspective
Kayhan Erciyes
ISBN 978-1-4665-7166-2

Data Privacy for the Smart Grid
Rebecca Herold and Christine Hertzog
ISBN 978-1-4665-7337-6

Generic and Energy-Efficient Context-Aware Mobile Sensing
Ozgur Yurur and Chi Harold Liu
ISBN 978-1-4987-0010-8

Machine-to-Machine Communications: Architectures, Technology, Standards, and Applications
Vojislav B. Misic and Jelena Misic
ISBN 978-1-4665-6123-6

Managing the PSTN Transformation: A Blueprint for a Successful Migration to IP-Based Networks
Sandra Dornheim
ISBN 978-1-4987-0103-7

MIMO Processing for 4G and Beyond: Fundamentals and Evolution
Edited by Mário Marques da Silva and Francisco A. Monteiro
ISBN 978-1-4665-9807-2

Mobile Evolution: Insights on Connectivity and Service
Sebastian Thalanany
ISBN 978-1-4822-2480-1

Network Innovation through OpenFlow and SDN: Principles and Design
Edited by Fei Hu
ISBN 978-1-4665-7209-6

Neural Networks for Applied Sciences and Engineering: From Fundamentals to Complex Pattern Recognition
Sandhya Samarasinghe
ISBN 978-0-8493-3375-0

Rare Earth Materials: Properties and Applications
A.R. Jha
ISBN 978-1-4665-6402-2

Requirements Engineering for Software and Systems, Second Edition
Phillip A. Laplante
ISBN 978-1-4665-6081-9

Security for Multihop Wireless Networks
Edited by Shafiullah Khan and Jaime Lloret Mauri
ISBN 9781466578036

Software Testing: A Craftsman's Approach, Fourth Edition
Paul C. Jorgensen
ISBN 978-1-46656068-0

The Future of Wireless Networks: Architectures, Protocols, and Services
Edited by Mohesen Guizani, Hsiao-Hwa Chen, and Chonggang Wang
ISBN 978-1-4822-2094-0

The Internet of Things in the Cloud: A Middleware Perspective
Honbo Zhou
ISBN 978-1-4398-9299-2

The State of the Art in Intrusion Prevention and Detection
Al-Sakib Khan Pathan
ISBN 978-1-4822-0351-6

ZigBee® Network Protocols and Applications
Edited by Chonggang Wang, Tao Jiang, and Qian Zhang
ISBN 978-1-4398-1601-1

AUERBACH PUBLICATIONS
www.auerbach-publications.com
To Order Call: 1-800-272-7737 • Fax: 1-800-374-3401 • E-mail: orders@crcpress.com

Multi-Carrier Communication Systems with Examples in MATLAB®

A New Perspective

Emad S. Hassan

CRC Press
Taylor & Francis Group
Boca Raton London New York

CRC Press is an imprint of the
Taylor & Francis Group, an **informa** business

CRC Press
Taylor & Francis Group
6000 Broken Sound Parkway NW, Suite 300
Boca Raton, FL 33487-2742

Printed on acid-free paper
Version Date: 20150812

International Standard Book Number-13: 978-1-4987-3532-2 (Hardback)

Visit the Taylor & Francis Web site at
http://www.taylorandfrancis.com

and the CRC Press Web site at
http://www.crcpress.com

Contents

PART II

CHAPTER 5   PERFORMANCE EVALUATION OF THE OFDM AND SC-FDE SYSTEMS USING CONTINUOUS PHASE MODULATION

PART III

CHAPTER 6   CHAOTIC INTERLEAVING SCHEME FOR THE CPM-OFDM AND CPM-SC-FDE SYSTEMS

PART IV

Preface

Next-generation wireless communication systems require higher data rate transmission in order to meet the higher demand of quality services and to support wideband wireless services such as high-definition TV, mobile videophones, video conferencing, and high-speed Internet access. Multi-carrier transmission techniques are attractive for wideband communications because they effectively transform the frequency-selective fading channel into a flat fading channel. Orthogonal frequency division multiplexing (OFDM) is one of the most well-known examples of multi-carrier transmission techniques. OFDM provides greater immunity to multipath fading and impulsive noise and eliminates the need for complicated equalizers. In addition, the combination of multiple-input-multiple-output (MIMO) signal processing techniques with OFDM is regarded as a promising solution to give a good performance and support high data rate transmission in next-generation wireless communication systems.

However, OFDM communication systems have two primary drawbacks. The first is the high sensitivity to carrier frequency offsets and phase noise. The second, and the major aspect we are concerned with, is the high peak-to-average power ratio (PAPR) of the transmitted signals. The high PAPR makes OFDM sensitive to nonlinear distortion caused by the transmitter power amplifier, hence leading to degradation in system performance. This book provides new solutions for these problems, and its main objective is to enhance the performance of multi-carrier systems.

A major attraction of the book is the presentation of MATLAB® simulations and the inclusion of MATLAB codes to help readers understand the topic under discussion and be able to carry out extensive simulations. The book is structured into eight chapters and broadly covers four parts, as follows:

- Part I (Chapters 1 through 4) starts with a detailed overview of multi-carrier systems such as OFDM, multi-carrier code division multiple access (MC-CDMA) and single-carrier frequency division multiple access (SC-FDMA) systems. We also explain the single-carrier scheme with frequency domain equalization (SC-FDE) scheme, which is another way to deal with the frequency-selective fading channel. This is followed by an overview of the PAPR problem in multi-carrier systems. A general overview of MIMO concepts and the MIMO-OFDM system model are also presented in this part of the book. We then study the selective mapping (SLM) scheme, which is one of the most popular PAPR reduction techniques proposed for MIMO-OFDM systems. To conclude this part, we propose a small overhead SLM (s-SLM) scheme for space-time block coded (STBC) MIMO-OFDM systems. This scheme improves the system bandwidth efficiency and achieves a significantly lower bit error rate (BER) than the individual SLM (i-SLM) and direct SLM (d-SLM) schemes. Also, the PAPR performance of the proposed s-SLM scheme is improved by using an unequal power distribution approach. This approach is based on assigning powers to the different subcarriers of OFDM using an unequal power distribution strategy.
- Part II (Chapter 5) starts with a study of the performance of the continuous phase modulation (CPM)-based OFDM (CPM-OFDM) system. The CPM is a new PAPR reduction technique in which the high PAPR OFDM signal is transformed to a constant envelope signal (i.e., 0 dB PAPR). We then propose a CPM-based single-carrier frequency domain equalization (CPM-SC-FDE) structure for broadband

wireless communication systems. Both the CPM-OFDM system and the proposed CPM-SC-FDE system are implemented with frequency domain equalization (FDE) to get high diversity gains over the frequency-selective multipath fading channels. Three types of frequency domain equalizers are considered: the zero forcing (ZF) equalizer, the regularized zero forcing (RZF) equalizer, and the minimum mean square error (MMSE) equalizer.

- In Part III (Chapter 6), we propose a chaotic interleaving scheme for both CPM-OFDM and the CPM-SC-FDE systems. Chaotic interleaving is used in these systems to generate permuted versions from the sample sequences to be transmitted, with low correlation among their samples and hence a better BER performance. A comparison between the proposed chaotic interleaving and the conventional block interleaving is also performed in this part.

- Part IV (Chapters 7 and 8) proposes different and efficient image transmission techniques over multi-carrier systems such as OFDM, MC-CDMA, and SC-FDMA. We begin by presenting a new approach for efficient image transmission over OFDM and MC-CDMA systems using chaotic interleaving. This approach transmits images over wireless channels efficiently, without posing significant constraints on the wireless communication system bandwidth and noise. The performance of the proposed approach is further improved by applying frequency domain equalization (FDE) at receiver.

We then study the performance of discrete cosine transform–based single-carrier frequency division multiple access (DCT-SC-FDMA) with image transmission. We also propose a CPM-based DCT-SC-FDMA structure for efficient image transmission. The structure presented has the advantages of excellent spectral energy compaction property of DCT-based SC-FDMA in addition to exploiting the channel frequency diversity and the power efficiency of CPM. Two types of transforms are considered and compared for performance evaluation of CPM-based SC-FDMA: discrete Fourier transform (DFT) and DCT. In addition, the performance of the proposed CPM-based DCT-SC-FDMA structure is compared with the conventional quadrature phase shift keying (QPSK)-based SC-FDMA system. Simulation experiments are performed using additive white Gaussian noise (AWGN) channel.

- Finally, MATLAB codes for all simulation experiments are included in Appendices A and B.

MATLAB® is a registered trademark of The MathWorks, Inc. For product information, please contact:

The MathWorks, Inc.
3 Apple Hill Drive
Natick, MA 01760-2098 USA
Tel: 508-647-7000
Fax: 508-647-7001
E-mail: info@mathworks.com
Web: www.mathworks.com

Acknowledgments

First and foremost, I am thankful to God, the most gracious most merciful, for helping me finish this work.

I express my sincere thanks to Professors Said E. El-Khamy, Moawad Dessouky, Sami El-Dolil, and Dr. Fathi E. Abd El-Samie. I am deeply indebted to them for valuable supervision, continuous encouragement, useful suggestions, and active help during the course of this work. I also extend my gratitude to Dr. Xu Zhu for her valuable technical discussions during my study at Liverpool University, United Kingdom.

Many thanks are extended to the authors of all journals and conference papers, articles, and books that have been consulted in writing this book. I extend my gratitude to all my past and current MSc and PhD students for their immense contributions to the knowledge in the area of image transmission over multi-carrier systems. Their contributions have undoubtedly enriched the content of this book.

Finally, I remain extremely grateful to my family who have continued to be supportive and provided needed encouragement. In particular, very special thanks go to my wife, Samah A. Ghorab, for her continuous patience and unconditional support that has enabled me to finally complete this challenging task. Her support has been fantastic.

Author

Dr. Emad S. Hassan earned his BSc (Honors), MSc, and PhD from the Electronics and Electrical Communications Engineering Department, Faculty of Electronic Engineering, Menoufia University, Egypt, in 2003, 2006, and 2010, respectively. In 2008, he joined the Communications Research Group at Liverpool University, Liverpool, United Kingdom, as a visiting researcher to complete his PhD research.

Dr. Hassan has been a full-time demonstrator (2003–2006) and assistant lecturer (2007–2010) at the Faculty of Electronic Engineering, Menoufia University. He was a visiting researcher at University of Liverpool, (2008–2009), a teaching assistant at the University of Liverpool (2008–2009), and a part-time lecturer at several private engineering universities in Egypt (2010–2011). He co-supervises many MSc and PhD students (2010–present). Currently, he is assistant professor at the Electronics and Electrical Communications Engineering Department, Faculty of Electronic Engineering, Menoufia University, Egypt.

Dr. Hassan is a reviewer for many international journals and conferences. He has been a Technical Program Committee member for several international conferences. He has published more than 50 scientific papers in national and international conference proceedings and journals. His current research areas of interest include image processing, digital communications, cooperative communication, cognitive radio networks, OFDM, SC-FDE, MIMO, and CPM-based systems.

List of Abbreviations

A/D	Analog-to-digital
ACI	Adjacent channel interference
AWGN	Additive white Gaussian noise
BER	Bit error rate
BLAST	Bell Labs space-time architecture
BPSK	Binary PSK
CCDF	Complementary cumulative distribution function
CDF	Cumulative distribution function
CPM	Continuous phase modulation
CPM-OFDM	Continuous phase modulation–based OFDM
CPM-SC-FDE CPM	Continuous phase modulation–based single-carrier frequency domain equalization
CSI	Channel state information
DAB	Digital audio broadcasting
D/A	Digital-to-analog
DFT	Discrete Fourier transform
DSL	Digital subscriber lines
d-SLM	Direct SLM

DVB-T	Terrestrial digital video broadcasting
FDE	Frequency domain equalization
FDM	Frequency division multiplexing
FFT	Fast Fourier transform
FIR	Finite impulse response
IBI	Interblock interference
IBO	Input power backoff
ICI	Intercarrier interference
IDFT	Inverse discrete Fourier transform
IFFT	Inverse fast Fourier transform
i.i.d.	Identically distributed
ISI	Intersymbol interference
i-SLM	Individual SLM
IQ	In phase and quadrature
MIMO	Multiple-input-multiple-output
MISO	Multiple-input-single-output
MMSE	Minimum mean square error
OFDM	Orthogonal frequency division multiplexing
PA	Power amplifier
PAM	Pulse amplitude modulation
PAPR	Peak-to-average power ratio
PLC	Power line communication
PTS	Partial transmit sequences
QAM	Quadrature amplitude modulation
QPSK	Quadrature phase shift keying
RMS	Root-mean-square
RZF	Regularized zero forcing
SC-FDE	Single-carrier frequency domain equalization
SI	Side information
SIMO	Single-input-multiple-output
SISO	Single-input-single-output
SLM	Selective mapping
SNR	Signal-to-noise ratio
S/P	Serial-to-parallel
SSPA	Solid-state power amplifier
s-SLM	Small overhead selective mapping
STC	Space-time coding
STBC	Space-time block coding

STF	Space-time-frequency
TWTA	Traveling-wave tube amplifier
WLANs	Wireless local area networks
WSSUS	Wide-sense stationary uncorrelated scattering
ZF	Zero forcing

List of Symbols

A	Input signal amplitude
A_{sat}	Input saturation level
$B^{(v)}$	Different phase sequences
B_{loss}	Percentage of bandwidth degradation
C	Scaling constant
C_n	Normalization constant
D	Number of diversity channels
E_s	Total transmitted power
f_k	Center frequency of the kth subcarrier
f_{nor}	Normalized cut-off frequency
G	Amplitude characteristics of the PA
G_0	Amplifier gain
$\mathbf{H}$	Channel matrix
$h(\tau)$	Channel impulse response
h	Modulation index
$\mathbf{H}[k]$	DFTs of $h(m)$
$h_{i,j}$	Fading coefficient from the jth transmit antenna to the ith receive antenna
IDFT{.}	IDFT operator
I_k	Real-valued data symbols
J	Oversampling factor
K	Number of subcarriers

L	Number of discrete paths
L_f	Length of filter impulse response
L_s	Number of sub-blocks
M	Number of constellation points
m	Number of bits per symbol
m_g	Multiplexing gain
M_s	Length of sub-block
$n(t)$	Additive noise
$n[i]$	Noise signal samples
N_0	Power spectral density
N_b	Number of samples per OFDM block
N_c	Channel samples
N_{DFT}	Number of DFT points
N_g	Number of guard samples
N_r	Number of receive antennas
N_{SI}	Number of required SI bits
N_t	Number of transmit antennas
P	Controls the sharpness of the saturation region
P_{avg}	Average power
P_d	Detection probability of the SI bits
P_e	BER of the MIMO-OFDM system
P_{peak}	Maximum envelope power
$q_k(t)$	Orthogonal subcarriers
$r(t)$	Received signal
R	Bit rate
$r[i]$	Received signal samples
R_c	Coding rate
$\tilde{s}(n)$	Soft estimate of $s(n)$
$s(n)$	Constant envelope sequence
$s(t)$	CPM-based signal
$s_I(n)$	Subscript I referring to interleaving process
$S_{in}(t)$	Input signal to the PA
$S_{out}(t)$	Output signal from the PA
T	OFDM symbol time
T_g	Guard time length
T_s	Symbol period
V	Number of different phase sequences
$W(m)$	Equalizer coefficients

W	Effective double-side bandwidth of the message signal
$x(t)$	OFDM transmitted signal
x^*	Complex conjugate of x
$x[i]$	Transmitted signal samples
$X[k]$	DFTs $x(m)$
$X_i^{(V)}$	Modified data blocks
X_k	Data symbol of kth subcarrier
$(.)^{\mathrm{T}}$	Matrix transposes
$\lceil x \rceil$	Smallest integer greater than or equal to x
$\Im\{X(k)\}$	Imaginary part of $\{X(k)\}$
$\Re\{X(k)\}$	Real part of $\{X(k)\}$
α	Factor larger than one
$\alpha_i = \|b_i\|$	Amplitude gain for the path from transmit antenna i to the receive antenna
γ_o	Average SNR per antenna
$\delta(n)$	Phase of the noise signal
Δ	PAPR loss compared to SISO-SLM scheme
Δf	Subcarrier spacing
η	Bandwidth efficiency of the CPM signal
η_{PA}	PA efficiency
θ	An arbitrary phase offset used to achieve CPM
θ_i	Input signal phase
σ^2_I	Variance of the data symbols
τ_l	Channel delay of the lth path
$\tau_{\max}$	Maximum propagation delay
$\phi(t)$	Phase of CPM signal
$\phi_a(t)$	Phase characteristics of the PA

PART I

1

INTRODUCTION

The goal of the third- and fourth-generation mobile networks is to provide high data rates and a wider range of services, such as voice communications, videoconferencing, and high-speed Internet access. A common challenge in designing a wireless system is to overcome the effects of the wireless channel, which is characterized as having multiple transmission paths and as being time varying. Orthogonal frequency division multiplexing (OFDM) has a promising future as a new technology in several next-generation wireless communication systems. The ability of OFDM systems to combat the effects of multipath propagation with a comparatively simple receiver structure has made it the modulation technique of choice for some of the most prominent wireless technologies, such as the IEEE 802.11 wireless local area networks (WLANs). It is also used in wireless broadcasting applications such as digital audio broadcasting and terrestrial digital video broadcasting (DVB-T) [1–3]. OFDM has also been implemented in wireline applications such as digital subscriber lines [4] and power line communication [5].

1.1 Orthogonal Frequency Division Multiplexing

One way to mitigate the frequency-selective fading seen in a wideband channel is to use a multi-carrier technique, which subdivides the channel into smaller sub-bands, or subcarriers. In conventional single-carrier systems, a single fade or interferer can cause the entire link to fail, but in multi-carrier systems, only a small percentage of the subcarriers will be affected. Error correction coding can then be used to correct the few erroneous subcarriers. OFDM is a multi-carrier technique which uses orthogonal subcarriers to convey information. In the frequency domain, since the bandwidth of a subcarrier is designed to be smaller than the coherence bandwidth, each subchannel is seen as a flat fading channel, which simplifies the channel equalization process. In the time domain, by splitting a high-rate data stream into a number of lower-rate data streams that are transmitted in parallel, OFDM resolves the problem of intersymbol interference in wideband communications [6].

1.1.1 OFDM Advantages

OFDM has the following advantages:

OFDM is an efficient way to deal with multipath effects. For a given channel delay spread, the implementation complexity is much lower than that of a conventional single-carrier system with a time domain equalizer.

Bandwidth efficiency is high since it uses overlapping orthogonal subcarriers in the frequency domain.

Modulation and demodulation are implemented using inverse discrete Fourier transform and discrete Fourier transform (DFT), respectively. The fast Fourier transform algorithm can be applied to make the overall system more efficient.

Capacity can be significantly increased by adapting the data rate per subcarrier according to the signal-to-noise ratio of that particular subcarrier.

OFDM is robust against narrowband interference because such interference affects only a small percentage of the subcarriers.

1.1.2 Problems Associated with OFDM

OFDM communication systems have two primary drawbacks. The first is the high sensitivity to carrier frequency offsets and phase noise. When there are frequency offsets in the subcarriers, the orthogonality among the subcarriers breaks and causes intercarrier interference. The second drawback is that the transmitted OFDM signal has large amplitude fluctuations, and therefore a high peak-to-average power ratio (PAPR). This high PAPR requires system components with a wide linear range to accommodate signal variations. Otherwise, nonlinear distortion, which results in loss of subcarrier orthogonality and hence degradation in the system performance, occurs.

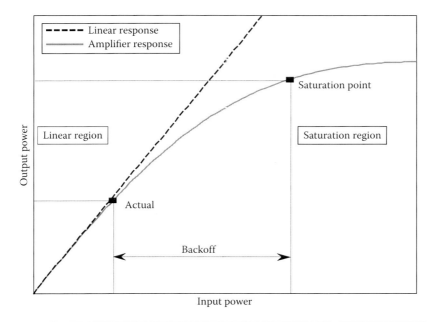

Figure 1.1 Power amplifier transfer function.

One nonlinear device is the transmitter power amplifier (PA). The PA has a limited linear region, beyond which it saturates to a maximum output level. Figure 1.1 shows the PA transfer function. The most efficient operating point is at the PA saturation point, but for signals with large PAPR, the operating point must shift to the left, keeping the amplification linear. The average input power is reduced, and consequently this technique is called input power backoff (IBO). To keep the peak power of the input signal less than or equal to the saturation input level, the IBO must be at least equal to the PAPR [7].

A high PAPR makes OFDM sensitive to the nonlinear distortion caused by the transmitter PA, leading to degradation in system performance. Therefore, to improve the OFDM system performance, we need to first solve the PAPR problem.

1.2 SC-FDE System

For broadband multipath channels, conventional time domain equalizers are impractical because of their complexity. Frequency domain equalization (FDE) is more practical for such channels. The single-carrier frequency domain equalization (SC-FDE) scheme is another way to deal with the frequency-selective fading channel. Its performance is similar to OFDM with essentially the same overall complexity [8–12]. Chapter 2 gives a detailed overview of this system.

1.3 MC-CDMA System

Combining OFDM and code division multiple access (CDMA) results in a multi-carrier CDMA (MC-CDMA) system [3,13,14]. The MC-CDMA system has received much research attention. However, it suffers from multiple access interference (MAI) in a multiuser setting, which decreases the overall bit error rate (BER) performance. Multiuser detection techniques have been introduced to mitigate MAI in order to improve system performance [15,16]. This book studies the efficient image transmission over MC-CDMA-based systems.

1.4 Single-Carrier Frequency Division Multiple Access System

The single-carrier frequency division multiple access (SC-FDMA) system is a multiple access scheme and is used as an uplink transmission in the long-term evolution (LTE) of cellular systems to make the mobile terminal power efficient. SC-FDMA is the multiuser version of SC-FDE, whereas OFDMA is the multiuser version of OFDM [17,18].

There are two variants of SC-FDMA, which differ in the manner in which the subcarriers are mapped to a particular user. They are the interleaved FDMA (IFDMA), which assigns equidistant subcarriers to each user, and the localized FDMA (LFDMA), which assigns contiguous subcarriers to a particular user. With respect to immunity to transmission errors, which determines throughput, IFDMA is robust against frequency-selective fading because its information is spread across the entire signal band. Therefore, it offers the advantage of frequency diversity [19–21].

1.4.1 Comparison between OFDM and SC-FDMA

Broadband wireless mobile communications suffer from multipath frequency-selective fading. OFDM, which is a multi-carrier communication technique, became accepted because of its robustness against frequency-selective fading channels, which are common in broadband mobile wireless communications [17].

SC-FDMA utilizes single-carrier modulation and frequency domain equalization. SC-FDMA is similar to the OFDMA system in performance and has essentially the same overall complexity [1,2,20]. Therefore, SC-FDMA is referred to as DFT-spread OFDMA. OFDMA is a multiple access scheme which is an extension of OFDM to accommodate multiple simultaneous users. OFDM/OFDMA technique is currently adopted in WLAN, WiMAX, and 3GPP LTE downlink systems. SC-FDMA is based on the same principle as SC-FDE; the only difference is SC-FDMA is for multiple users, whereas SC-FDE is a single-user modulation scheme [18].

On the receiver side, frequency domain equalization is done in OFDMA on a per-subcarrier basis, whereas in SC-FDMA it is done by using a complex equalizer used for all the subcarriers together. The receiver structure is therefore complex in SC-FDMA compared to OFDMA. For broadband multipath channels, conventional time domain equalizers are impractical, for complexity reasons.

Although OFDM and OFDMA have advantages, they suffer from a number of drawbacks including high PAPR, a need for an adaptive or coded scheme to overcome spectral nulls in the channel, and high sensitivity to frequency offset. On the other hand, SC-FDMA has an advantage over OFDMA in that the SC-FDMA signal has lower PAPR because of its inherent single-carrier structure. The parallel transmission of subcarriers in OFDMA gives rise to high PAPR, and the sequential transmission of subcarriers in SC-FDMA gives low PAPR. The lower PAPR greatly benefits the mobile terminal in terms of transmit power efficiency and manufacturing cost and allows use of simple power amplifiers that reduces power consumption [22]. Hence, SC-FDMA has gained much attention as an attractive alternative to OFDMA, especially in the uplink communications where the receiver is placed in the base station and the transmitter is at the mobile station, since power efficiency and complexity are more important in mobile stations than in base stations.

1.5 Image Transmission

Due to the importance of images in modern telecommunications, this book proposes various and efficient image transmission systems over multi-carrier systems. First we must define the image formation, and then the image definition. Finally, we define the peak signal-to-noise ratio (PSNR) and the application which needs image transmission.

1.5.1 Image Formation

Image formation is the process by which a visual scene is transformed into a form that can be processed. This image formation process is illustrated in Figure 1.2. In this figure, an object $f(x_1, y_1)$ in the coordinate system (x_1, y_1), which is referred to as the object plane, is illuminated by a source of radiant energy. The transmitted radiant energy propagates through space. An image formation system intercepts the propagating radiant energy and transforms it in such a manner that in the coordinate system (x, y), which is referred to as the image plane, an image is formed. The process of image formation is light dependent and thus light perception is of great interest in image transmission [23,24].

1.5.2 Image Definition

An image may be defined as a two-dimensional (2-D) function, $f(x, y)$, where x and y are spatial (plane) coordinates, and the amplitude of f at any pair of coordinates (x, y) is called the intensity or gray level of the image at that point. When x, y, and the amplitude values of f are all finite, discrete quantities, we call the image a digital image. Note that a digital image is composed of a finite number of elements, each of which has a particular location and value. These elements are referred to as picture elements, image elements, and pixels.

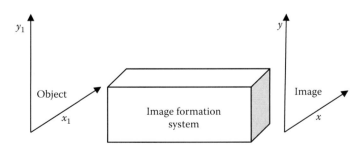

Figure 1.2   Linear imaging system model.

Pixel is the term most widely used to denote the elements of a digital image [25]. Then an image can be defined as a 2-D signal (analog or digital), that contains intensity (grayscale), or color information arranged along *x* and *y* spatial axes [26].

1.5.2.1 Analog Image An analog image can be mathematically represented as a continuous range of values representing position and intensity [26].

1.5.2.2 Digital Image A digital image is restricted in both its spatial coordinates and allowed intensities. The positions and intensities are represented with discrete values, or elements. The discrete elements that make up a digital image are called picture elements, or pixels. Each pixel consists of one or more bits of information that represent either the brightness or brightness and color. Matrices are perfect tools for mapping, representing digital images. For example, an image of 600 pixels height and 800 pixels width can be represented as a 600 × 800 matrix (600 rows and 800 columns). Each element of the matrix (pixel) is used to represent intensity [26].

The acquisition of a digital image is performed in three steps:

1. *Sampling*: Obtaining intensity readings at evenly spaced locations in both the *x* and *y* directions. This is performed by placing an evenly spaced grid over the analog image. The readings obtained at these locations are the intensities of pixels.
2. *Quantizing*: The sampled values of intensity are quantized to arrive at a signal that is discrete in both positions and amplitudes. This signal represents the image.
3. *Encoding*: The conversion of the data into a binary form (Figure 1.3) [26].

A scanner or a digital camera is commonly used for digitizing images. They contain sensor arrays that react to different intensities and wavelengths. The chip inside the camera is a digital image sensor that

Figure 1.3 Conversion from an analog image to a digital image.

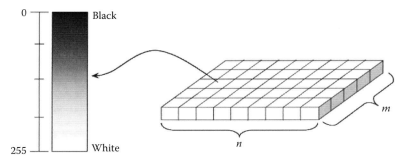

Figure 1.4   Mapping of pixel values into the grayscale.

consists of millions of individual elements capable of capturing light. The light-sensitive elements transform light energy to voltage values based on the intensity of the light. The voltage values are then converted to digital data by an analog-to-digital converter chip. The digital numbers corresponding to the voltage values for each element combine to create the tonal values of the image (Figure 1.4).

To digitize a grayscale image, we look at the overall intensity levels of the sensed light and record them as a function of position. To digitize a color image, the intensities of each of the three primary color components—red, green, and blue—must be detectable from the incoming light. One way to accomplish this is to filter the light sensed by the sensor so that it lies within the wavelength range of a specific color. Thus, we can detect the intensity of that specific color for that specific sample location.

To quantize a color image, consider a computer imaging system that utilizes 24-bit color. For the 24-bit color system, each of the three primary color intensities is allowed one byte of storage per pixel for a total of three bytes per pixel (Figure 1.5). Each color has an allowed numerical range from 0 to 255; for example, 0 means no red and 255 means all red. The combinations that can be made with 256 levels for each of the

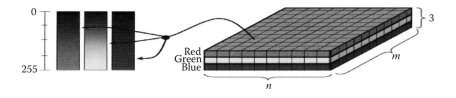

Figure 1.5   3-D array representing the 24-bit color system.

three primary colors amounts to over 16 million distinct colors ranging from black (R, G, B) = (0, 0, 0) to white (R, G, B) = (255, 255, 255). Most computers store the information of color digital images in three-dimensional (3-D) arrays. The first two indices in this array specify the row and column of the pixel and the third index specifies the color plane, where 1 is red, 2 is green, and 3 is blue [26,27].

1.5.3 Peak Signal-to-Noise Ratio

PSNR is used to measure the quality of the reconstructed images at the receiver. It is the ratio between the maximum possible power of a signal and the power of corrupting noise that affects the fidelity of this signal. Because many signals have a very wide dynamic range, PSNR is usually expressed in terms of the logarithmic decibel scale. PSNR is defined as follows [19]:

$$PSNR = 10\log_{10}\left(\frac{\max_f^2}{MSE}\right) \tag{1.1}$$

where $\max_f$ is the maximum possible pixel value of an image f. For 8-bit pixels, $\max_f = 255$. MSE is the mean square error. For an $N \times N$ monochrome image, it is defined as

$$MSE = \frac{\sum\left[f(i,j) - \hat{f}(i,j)\right]^2}{N^2} \tag{1.2}$$

where
$f(i, j)$ is the source image
$\hat{f}(i, j)$ is the reconstructed image

To evaluate the performance and efficiency of these systems, the 256 × 256 cameraman image is used.

1.5.4 Applications That Need Image Transmission

Image transmission is used in many applications today, including teleconferencing, computer communication, facsimile transmission, telemedicine, teleradiology, e-learning, nuclear medicine, broadcast television, remote sensing via satellite, military applications via aircraft, and radar and sonar.

The rapid development of the Internet, videoconferencing, telephony, and the World Wide Web plays an important role in distance education and healthcare, collaboration among the users (e.g., teleconferencing) especially in telemedicine, transfer of images of high resolution are vital [27].

Telemedicine means "distant care." It can be broadly defined as the use of information technology to deliver medical services between distant sites. It utilizes information and telecommunications technology to transfer medical information for diagnosis, therapy, and education. It can include the transfer of basic patient information over computer networks (medical informatics), the transfer of images such as radiographs, computerized tomography scans, magnetic resonance imaging, ultrasound studies, pathology images, video images of endoscopic or other procedures, patient interviews and examinations, consultations with medical specialists, and healthcare educational activities.

Telemedicine services enable healthcare systems to enhance coverage and quality of care using the available broad bandwidth communication capabilities. Developments in the field of telecommunication and computers have provided the capability of videoconferencing. In short, telemedicine draws on the technologies of videoconferencing and requires the integration of information from biomedical instruments used in clinical settings.

Nuclear medicine is a medical specialty involving the application of radioactive substances in the diagnosis and treatment of disease.

In nuclear medicine procedures, radionuclides are combined with other elements to form chemical compounds or are combined with existing pharmaceutical compounds to form radiopharmaceuticals. These radiopharmaceuticals, once administered to the patient, can localize to specific organs or cellular receptors. This property of radiopharmaceuticals allows nuclear medicine the ability to image the extent of a disease process in the body based on cellular function and physiology, rather than relying on physical changes in the tissue anatomy. In some diseases, nuclear medicine studies can identify medical problems at an earlier stage than can other diagnostic tests. Nuclear medicine, in a sense, is "radiology done inside out" or "endo-radiology," because it records radiation emitting from within the body rather than radiation that is generated by external sources such as x-rays.

In the future, nuclear medicine may provide added momentum to the field of molecular medicine. As the understanding of biological processes in the cells of living organism expands, specific probes can be developed to allow visualization, characterization, and quantification of biologic processes at the cellular and subcellular levels. Nuclear medicine is a promising specialty for adapting to the new discipline of molecular medicine because of its emphasis on function and its utilization of imaging agents that are specific for a particular disease process [27].

E-learning is essentially the computer and network-enabled transfer of skills and knowledge. E-learning applications and processes include web-based learning, computer-based learning, virtual education opportunities, and digital collaboration. Content is delivered via the Internet, intranet/extranet, audio or video tape, satellite TV, and CD-ROM. It can be self-paced or instructor-led and includes media in the form of text, image, animation, and streaming video and audio [25].

1.6 Book Objectives and Contributions

- *Overview of the OFDM system basics*: The basic principles of the OFDM system are described. We also illustrate its similarities with and differences from the SC-FDE system.
- *Overview of the PAPR problem in multi-carrier systems*: A detailed overview of the PAPR problem is given followed by a study of the effects of PA nonlinearity on OFDM system performance.
- *Study of the MIMO space-time block coding (STBC) principles*: The basic principles of MIMO systems are briefly explained. Then we present the MIMO-OFDM system model.
- *Proposal of the small overhead selective mapping (s-SLM) scheme*: The SLM scheme is one of the most popular PAPR reduction techniques proposed for MIMO-OFDM systems. We propose a new s-SLM scheme for STBC MIMO-OFDM systems, which improves system bandwidth efficiency and achieves a significantly lower BER than the individual (i-SLM) and direct (d-SLM) schemes.
- *Study of the performance of the continuous phase modulation–based OFDM (CPM-OFDM) system*: The CPM is a new PAPR

reduction technique in which the high PAPR OFDM signal is transformed to a constant envelope signal (0 dB PAPR).

- *Proposal of a CPM-SC-FDE structure for broadband wireless communication systems*: We propose a new structure called the CPM-SC-FDE in which we use CPM to improve the performance of the SC-FDE system. Both the CPM-OFDM system and the proposed CPM-SC-FDE system are implemented with FDE to obtain high diversity gains over frequency-selective multipath fading channels.

- *Proposal of a chaotic interleaving scheme for the CPM-OFDM system and the CPM-SC-FDE system*: Chaotic interleaving is used to generate permuted versions from the sample sequences to be transmitted, with low correlation among sample sequences after chaotic interleaving and hence a better BER performance.

- *Proposal of efficient image transmission techniques over multi-carrier systems such as OFDM, multi-carrier code division multiple access (MC-CDMA), and SC-FDMA*: We start by proposing a new approach for efficient image transmission over OFDM and MC-CDMA systems using chaotic interleaving. We then study the performance of discrete cosine transform based single carrier frequency division multiple access (DCT-SC-FDMA) with image transmission. We also propose a CPM-based DCT-SC-FDMA structure for efficient image transmission.

1.7 Chapter Outlines

Chapter 2 gives a general overview of the OFDM system. We provide an overview of the basic principles of OFDM and then present the SC-FDE system, followed by an overview of the PAPR problem. Finally, we study the effects of the PA nonlinearity on the OFDM system performance and outline the PAPR reduction techniques.

Chapter 3 investigates the MIMO techniques for the OFDM system. We give a general overview of the MIMO concepts and then introduce the MIMO-OFDM system model.

Chapter 4 proposes an s-SLM scheme. This scheme improves the system bandwidth efficiency and achieves a significantly lower BER than the i-SLM and the d-SLM schemes. In addition, approximate expressions for the complementary cumulative distribution function (CCDF) of the PAPR and the average BER of the proposed s-SLM scheme are derived. The results show that the proposed s-SLM scheme improves the detection probability of the side information (SI) bits and hence gives a better performance than the i-SLM and the d-SLM schemes.

Chapter 5 studies the performance of the CPM-OFDM system. We also propose a CPM-SC-FDE structure for broadband wireless communication systems which combines the advantages of the frequency diversity and the low complexity of the SC-FDE with the power efficiency of the CPM. Both the CPM-OFDM system and the proposed CPM-SC-FDE system are implemented with FDE to obtain high diversity gains over frequency-selective multipath fading channels. Simulation experiments are performed for a variety of multipath fading channels, the results of which show that the performance of the CPM-based systems with multipath fading is better than with single-path fading. The performance over multipath channels is at least 5 and 12 dB better than the performance over a single path channel for the CPM-OFDM system and the CPM-SC-FDE system, respectively.

Chapter 6 proposes a chaotic interleaving scheme for the CPM-OFDM system and the CPM-SC-FDE system. The proposed CPM-SC-FDE system with chaotic interleaving combines the advantages of the frequency diversity, the low complexity, and the high power efficiency of the CPM-SC-FDE system with performance improvements due to chaotic interleaving. The BER performance of the CPM-SC-FDE system with and without chaotic interleaving is evaluated by computer simulations. Also, a comparison between chaotic interleaving and block interleaving is performed. Simulation results show that the proposed chaotic interleaving scheme can greatly improve the performance of the CPM-OFDM system and the CPM-SC-FDE system. Furthermore, the results show that this scheme outperforms the traditional block interleaving scheme in both systems. The results also show that using chaotic interleaving with CPM-OFDM and CPM-SC-FDE systems provides a good trade-off between system performance and bandwidth efficiency.

Chapter 7 proposes a new approach for efficient image transmission over the multi-carrier OFDM–based system and the

MC-CDMA-based system using chaotic interleaving. The chaotic interleaving scheme based on the Baker map is applied on the image data prior to transmission. The proposed approach transmits images over wireless channels efficiently, without posing significant constraints on the wireless communication system bandwidth and noise. The performance of the proposed approach is further improved by applying FDE at receiver. Two types of frequency domain equalizers are considered and compared for performance evaluation of the proposed systems: the zero forcing (ZF) equalizer and the linear minimum mean square error (LMMSE) equalizer. Several experiments are carried out to test the performance of the image transmission with different sizes over the OFDM- and MC-CDMA-based systems.

Chapter 8 studies image transmission over SC-FDMA-based systems. The performance of two different structures—namely, the DFT-based SC-FDMA and the DCT-based SC-FDMA—are studied in order to select the proper technique for efficient image transmission. In this chapter, we also use the chaotic interleaving scheme with both SC-FDMA structures for efficient image transmission. Simulation of both structures using MATLAB® is performed and the experimental results show that the DCT-based SC-FDMA structure achieves higher PSNR values than the DFT-based SC-FDMA structure due to its excellent spectral energy compaction property. In addition, it uses basic arithmetic rather than the complex arithmetic used in the DFT-based SC-FDMA system. The results show that the PSNR values are enhanced by applying chaotic interleaving scheme in both structures.

In this chapter, we also study the performance of CPM-based DCT-SC-FDMA with image transmission. The proposed structure combines the advantages of excellent spectral energy compaction property of DCT-based SC-FDMA, in addition to exploiting the channel frequency diversity and the power efficiency of CPM. In addition, the performance of the proposed CPM-based DCT-SC-FDMA structure is compared with the conventional quadrature phase shift keying (QPSK)-based SC-FDMA system. Simulation experiments are performed using additive white Gaussian noise channel. The results show that the CPM-based DCT-SC-FDMA structure increases the transmission efficiency, provides better performance, and achieves higher PSNR values in the received images compared to conventional QPSK-based SC-FDMA systems.

2

BASIC PRINCIPLES OF MULTI-CARRIER COMMUNICATION SYSTEMS

2.1 Introduction

The basic principle of multi-carrier systems is to split a high-data-rate sequence into a number of lower-rate sequences that are transmitted simultaneously over a number of subcarriers. Because the symbol duration is increased for the low-rate parallel subcarriers, the relative amount of dispersion in time caused by multipath delay spread decreases. The intersymbol interference (ISI) is eliminated by introducing a guard interval at the start of each orthogonal frequency division multiplexing (OFDM) symbol. In the guard interval, an OFDM symbol is cyclically extended to avoid the intercarrier interference (ICI). Thus, a highly frequency-selective channel is transformed into a large set of individual flat fading, non-frequency-selective narrowband channels [1].

In this chapter, we first give an overview of the OFDM system principles. In Section 2.3, we present the single-carrier frequency domain equalization (SC-FDE) system. This is followed by an overview of the MC-CDMA system in Section 2.4. The peak-to-average power ratio (PAPR) problem is discussed in Section 2.5, and finally in Section 2.6, we study the effects of the PA nonlinearity on the OFDM system performance and outline the PAPR reduction techniques.

2.2 Basic Principles of OFDM

OFDM [1–4] is a special case of multi-carrier transmission, where a single data stream is transmitted over a number of closely spaced orthogonal subcarriers. The data stream is divided into several parallel data streams or subcarriers, one for each subcarrier. In the conventional single-carrier system, a single fade or interferer can cause the entire link to fail, but in a multi-carrier system, only a small percentage of the subcarriers will be affected. Error correction coding can then be used to correct the few erroneous subcarriers.

In the conventional frequency division multiplexing (FDM), the total signal frequency band is divided into nonoverlapping frequency subchannels. Each subchannel is modulated with a separate symbol, and then the subchannels are frequency-multiplexed. Spectral overlap of channels should be avoided to eliminate the ICI. However, this leads to inefficient use of the available spectrum, which in turn reduces the bandwidth efficiency. To improve the bandwidth efficiency, OFDM uses K overlapping subcarriers, whose orthogonality removes the interchannel interference with K representing the number of subcarriers.

Figure 2.1 illustrates the difference between a conventional nonoverlapping multi-carrier technique such as FDM and an overlapping multi-carrier modulation technique such as OFDM. By using the overlapping multi-carrier modulation technique, we save almost 50% of the bandwidth [2]. To realize this technique, however, we must achieve orthogonality between the different subcarriers. The term "OFDM" is derived from the fact that the digital (or analog) data is sent using several subcarriers, each having a different frequency, that are orthogonal to each other.

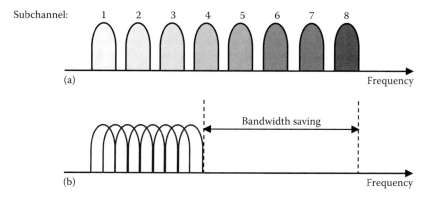

Figure 2.1 Representation of OFDM: (a) conventional multi-carrier technique (FDM) and (b) orthogonal multi-carrier modulation technique (OFDM).

2.2.1 *Orthogonality*

Signals are orthogonal if they are mutually independent of each other. Orthogonality is a property that allows multiple information signals to be transmitted perfectly over a common channel and detected without interference. Loss of orthogonality results in blurring between these information signals and in degradation of the communication channel. In the frequency domain, most FDM systems are orthogonal, as each of the separate transmitted signals are well spaced out in frequency to prevent interference. The term OFDM has been reserved for a special form of FDM. The subcarriers in an OFDM system are spaced as close as possible, but orthogonality is maintained.

OFDM achieves orthogonality in the frequency domain by allocating each of the separate information signals to a different subcarrier. The baseband frequency of each subcarrier is chosen to be an integer multiple of the symbol time reciprocal, resulting in all subcarriers having an integer number of cycles per symbol. As a consequence, the subcarriers are orthogonal to each other.

Figure 2.2 shows the construction of an OFDM signal with five subcarriers, where (1-a), (2-a), (3-a) (4-a), and (5-a) show individual subcarriers, with 1, 2, 3, 4, and 5 cycles per symbol, respectively. The phase on all these subcarriers is zero. Note that each subcarrier has an integer number of cycles per symbol, making them cyclic. Parts (1-b), (2-b), (3-b), (4-b), and (5-b) show the Fourier transform of the time waveforms in (1-a), (2-a), (3-a), (4-a), and (5-a), respectively. Parts (6-a) and (6-b) show the result for the summation of the five subcarriers.

Another way to view the orthogonality property of an OFDM signal is to look at its spectrum as shown in Figure 2.3. In the frequency domain, each OFDM subcarrier has a sinc, $\sin(x)/x$, frequency response. The *sinc* shape has a narrow main lobe, with many side lobes that decay slowly with the magnitude of the frequency difference away from the center. Each carrier has a peak at the center frequency and nulls evenly spaced with a frequency gap equal to the carrier spacing. The orthogonal nature of the transmission is maintained because the peak of each subcarrier corresponds to the nulls of all other subcarriers.

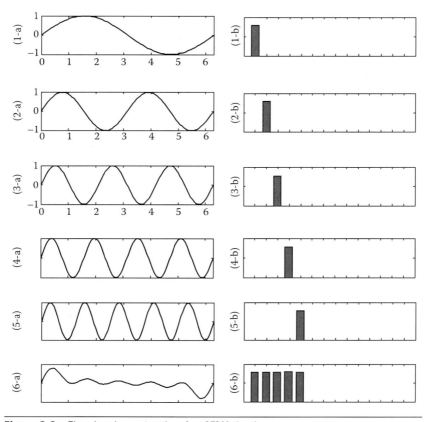

Figure 2.2 Time domain construction of an OFDM signal.

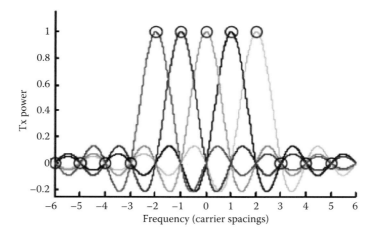

Figure 2.3 Frequency response of the subcarriers in a 5-tone OFDM signal.

2.2.2 Guard Time and Cyclic Prefix Extension

It is important in designing a wireless system to overcome the effects of the wireless channel. Multipath propagation is caused by the radio transmission signal reflected off objects in the propagation environment, such as walls, buildings, mountains, etc. These multiple signals arrive at the receiver at different times due to variations in transmission distances, as shown in Figure 2.4.

The received signal is the sum of several versions of the transmitted signal with varying delay and attenuation, and so ISI occurs. The severity of the ISI depends on the symbol period, T_s, relative to the channel maximum propagation delay, τ_{max}. To show this, let the symbol period be $T_s = 0.5$ µs and the maximum propagation delay of the channel be $\tau_{max} = 5$ µs, thus the ISI spans $\tau_{max}/T_s = 10$ symbols. Such ISI must be corrected at the receiver in order to provide reliable communication. The traditional approach to combating the ISI is the use of time domain equalizers [28,29]. However, time domain equalizers are typically impractical due to their high complexity, which grows with the ISI length.

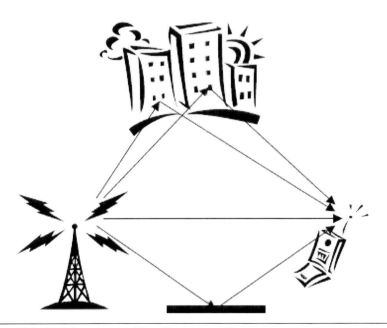

Figure 2.4   A wireless channel with multipath propagation.

Now, let us see how OFDM solves multipath problems such as ISI and ICI. In OFDM systems, the bandwidth is broken up into K subcarriers, resulting in a symbol rate that is K times lower than the single-carrier transmission. This low symbol rate makes OFDM naturally resistant to the effect of the ISI caused by multipath propagation. In other words, the OFDM symbol time T is thus K times longer than the original symbol time T_s (i.e., $T = KT_s$). To eliminate the ISI almost completely, a guard time is introduced for each OFDM symbol. The guard time is chosen larger than the expected delay spread so that the multipath versions of a symbol do not interfere with the next symbol. In that case, however, the problem of the ICI would arise. The ICI is a crosstalk between different subcarriers, which means that they are no longer orthogonal, as shown in Figure 2.5.

To eliminate the ICI, the OFDM symbol is cyclically extended in the guard time as shown in Figures 2.6 and 2.7. This ensures that the delayed replicas of the OFDM symbol always have an integer number of cycles within the fast Fourier transform (FFT) interval, as long as the delay is smaller than the guard time. As a result, multipath signals with delays smaller than the guard time cannot cause ICI [30,31].

The total OFDM symbol time is $T + T_g$, where T_g is the guard time length and T is the size of the IFFT used to generate the OFDM

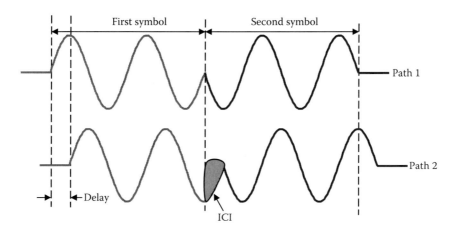

Figure 2.5   The ICI due to multipath propagation.

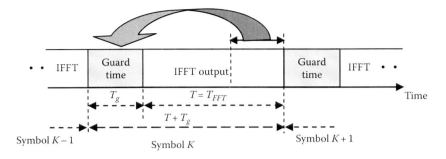

Figure 2.6 OFDM symbol with cyclic extension.

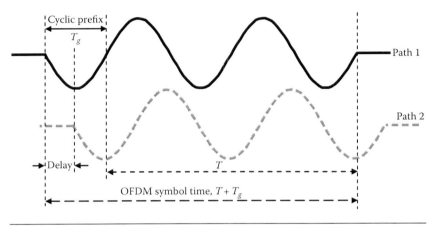

Figure 2.7 Example of an OFDM signal in a two-ray multipath channel.

signal. The main drawback of this principle is a slight loss of effective transmitted power, as the redundant guard time must be transmitted. Usually, the guard time is selected to have a length one-tenth to one quarter of the symbol time [2].

2.2.3 Time Domain Representation of the OFDM Signal

Instead of transmitting data symbols serially, OFDM sends K symbols as a block. Thus the OFDM block time, T, is K times longer than the original symbol time, T_s (i.e., $T = KT_s$). Suppose a block of K data symbols $\mathbf{X} = \{X_k, k = 0, 1,\ldots, K - 1\}$ is formed with each symbol modulating one of a set of subcarriers. The K subcarriers are chosen to be orthogonal, that is, $f_k = k\Delta f$, where $\Delta f = 1/T = 1/KT_s$.

Therefore, the complex envelope of the transmitted OFDM signals can be written as [30,31]

$$x(t) = \frac{1}{\sqrt{K}} \sum_{k=0}^{K-1} X_k \cdot e^{j2\pi f_k t}, \quad 0 \le t < T \tag{2.1}$$

where

X_k are the data symbols

$\exp(j2\pi f_k t)$ represents the subcarriers

$f_k = k/T$ is the center frequency of the kth subcarrier

Using OFDM and properly selecting the subcarrier spacing, Δf, the wideband frequency-selective channel is converted into K narrowband frequency-nonselective channels. The OFDM modulation can be optimized by sending more bits within the subcarrier with high SNR and fewer bits within the subcarrier with low SNR. This technique is known as the bit loading technique.

As explained in Section 2.2.2, the use of the cyclic prefix during the guard time results in an ISI-free operation. During the OFDM block time ($0 \le t < T$), the waveform is given by equation (2.1). The guard time is defined during $-T_g \le t < 0$. To transmit a cyclic prefix, the last T_g interval of the block is transmitted during the guard time as follows:

$$x(t) = \frac{1}{\sqrt{K}} \sum_{k=0}^{K-1} X_k \cdot e^{j2\pi f_k(T+t)}$$

$$= \frac{1}{\sqrt{K}} \sum_{k=0}^{K-1} X_k \cdot e^{j2\pi f_k t} \cdot e^{j2\pi k}, \quad -T_g \le t < 0$$

$$= \frac{1}{\sqrt{K}} \sum_{k=0}^{K-1} X_k \cdot e^{j2\pi f_k t} \tag{2.2}$$

This simplification is made due to the periodicity of the signal. Thus, the OFDM signal having a guard time with cyclic prefix is given as

$$x(t) = \frac{1}{\sqrt{K}} \sum_{k=0}^{K-1} X_k \cdot e^{j2\pi f_k t}, \quad T_g \le t < T \tag{2.3}$$

At the receiver side, the received signal is given by

$$r(t) = x(t) * h(\tau) + n(t)$$

$$= \int_{-\infty}^{\infty} h(\tau)x(t-\tau)d\tau + n(t) = \int_{0}^{\tau_{max}} h(\tau)x(t-\tau)d\tau + n(t) \quad (2.4)$$

where $h(\tau)$ is the channel impulse response with $0 \leq \tau \leq \tau_{max}$, τ_{max} is the channel maximum propagation delay, and $n(t)$ is the additive noise.

So far, two of the main OFDM advantages have been explained: the elimination of ISI and the ability to optimize the modulation with the bit loading technique. The third advantage of OFDM is that the modulation and demodulation can be performed in the discrete time domain using inverse direct Fourier transform (IDFT) and DFT, respectively.

2.2.4 Discrete Time Domain Representation of the OFDM Signal

By sampling $x(t)$, $h(\tau)$, and $r(t)$ at a sampling rate $f_s = JK/T$ sample/s, where $J \geq 1$ is the oversampling factor, the transmitted signal samples of equation (2.3) are given as

$$x[i] = x(t)\big|_{t=iT_{sa}} = \frac{1}{\sqrt{K}} \sum_{k=0}^{K-1} X_e \cdot e^{j2\pi ki/JK}, \quad N_g \leq i < N_b \quad (2.5)$$

where
$T_{sa} = 1/f_s$ is the sampling time
N_g is the number of guard samples ($N_g \leq T_g/T_{sa}$)
N_b is the number of samples per OFDM block ($N_b = N_{DFT} = JK$)

The received signal samples are expressed by the linear convolution sum as follows [32,33]:

$$r[i] = \sum_{m=0}^{N_c-1} h[m]x[i-m] + n[i], \quad N_g \leq i < N_b \quad (2.6)$$

where
N_c is the length of channel impulse response ($N_c \leq \tau_{max}/T_{sa}$) and $N_c \leq N_g$
$n[i]$ are the samples of the noise signal $n(t)$

The cyclic prefix samples are then removed:

$$r[i] = \sum_{m=0}^{N_c-1} h[m]x[i-m] + n[i], \quad 0 \leq i < N_b \quad (2.7)$$

The transmission of a cyclic prefix during the guard time makes the linear convolution given in equation (2.7) equivalent to a circular convolution [32,33]. The equivalent circular convolution can be written as

$$r[i] = \text{IDFT}\{H[k]X[k]\} + n[i]$$

$$= \frac{1}{N_{DFT}} \sum_{k=0}^{N_{DFT}-1} H[k]X[k] \cdot e^{j2\pi ki/N_{DFT}} + n[i] \quad 0 \leq i < N_b \quad (2.8)$$

where
   IDFT{·} is the IDFT operator
   $H[k]$ and $X[k]$ are the DFTs of $h(m)$ and $x(m)$, respectively

Figure 2.8 is a block diagram representing the calculation of equation (2.8). It is clear that the transmitted samples $x[i]$ can be reconstructed simply by passing the received samples $r[i]$ through the inverse channel.

The inverse channel structure in Figure 2.8 corrects the distortion caused by the channel in the frequency domain, and is therefore called a frequency domain equalizer [34,35]. Such an equalizer can be used only when the effect of the channel is a circular convolution, as in the OFDM case. We can conclude that OFDM does not eliminate

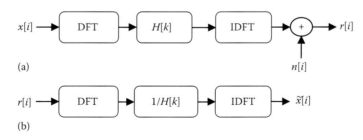

(a)

(b)

Figure 2.8   The effect of the channel is a circular convolution. (a) Channel and (b) inverse channel.

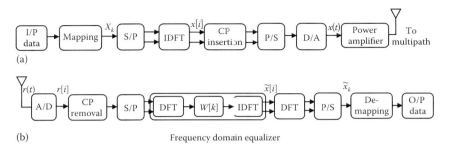

(a)

(b)   Frequency domain equalizer

Figure 2.9   OFDM system block diagram. (a) Transmitter and (b) receiver.

the equalization process associated with conventional single-car-rier systems; rather, it converts the problem to a frequency domain equalization.

2.2.5 OFDM System Block Diagram

Figure 2.9 is the block diagram of a typical OFDM transceiver. First, the input data symbols are mapped to the data symbols X_k. Then, the symbols are serial-to-parallel (S/P) converted and pro-cessed by the IDFT. The cyclic prefix is added and the signal sam-ples, $x[i]$, are passed through the digital-to-analog (D/A) converter to obtain the continuous-time OFDM signal $x(t)$. Finally, the sig-nal is amplified and transmitted. The receiver performs the reverse operation of the transmitter.

2.2.5.1 Modulation

The data allocated to each symbol depends on the used modulation scheme and the number of subcarri-ers. A modulation scheme is a mapping of data words to a real (In phase) and imaginary (In-quadrature) constellation, known as an IQ constellation. For example, 64-QAM (quadrature amplitude modulation) has 64 IQ points in the constellation, constructed in a square with eight evenly spaced columns in the real axis and eight rows in the imaginary axis. The num-ber of bits that can be transferred using a single symbol cor-responds to $m = \log_2(M)$, where M is the number of points in

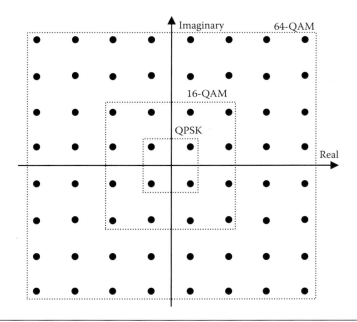

Figure 2.10   QPSK, 16-QAM, and 64-QAM constellation.

the constellation. Figure 2.10 shows quadrature phase shift keying (QPSK), 16-QAM, and 64-QAM constellation.

Increasing the number of points in the constellation does not change the bandwidth of the transmission, thus using a modulation scheme with a large number of constellation points will improve the spectral efficiency. The spectral efficiency of a channel is a measure of the number of bits transferred per second for each Hz. For example, 64-QAM has a spectral efficiency of 6 bps/Hz compared with only 1 bps/Hz for binary PSK (BPSK). However, the greater the number of points in the modulation constellation, the harder they are to resolve at the receiver. As the IQ locations become spaced close together, it only requires a small amount of noise to cause errors in the transmission. This results in a direct trade-off between noise tolerance and the spectral efficiency of the modulation scheme.

2.2.5.2 Serial to Parallel Conversion   Data to be transmitted is typically in the form of a serial data stream. Hence, an S/P conversion stage is needed to convert the input serial bit stream to the data to be transmitted in each OFDM symbol, as shown in Figure 2.11.

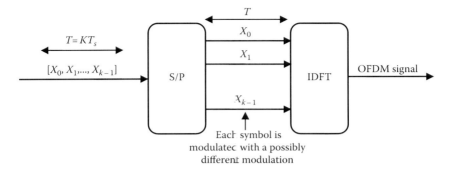

Figure 2.11   Generation of an OFDM signal using S/P conversion.

As explained earlier, the data allocated to each symbol depends on the modulation scheme used and the number of subcarriers. For adaptive modulation schemes, the modulation scheme used on each subcarrier can vary, and so the number of bits per subcarrier also varies. As a result, the S/P conversion stage involves filling the data payload for each subcarrier. At the receiver, the reverse process takes place, with the data from the subcarriers being converted back to the original serial data stream.

2.3 Basic Principles of the SC-FDE System

For broadband multipath channels, conventional time domain equalizers are impractical due to their complexity. FDE is more practical for such channels. The SC-FDE scheme is another way to deal with the frequency-selective fading channel. It has a performance similar to OFDM with essentially the same overall complexity [9–12,35]. Figure 2.12 shows the block diagram of the SC-FDE and compares it with that of the OFDM system [9].

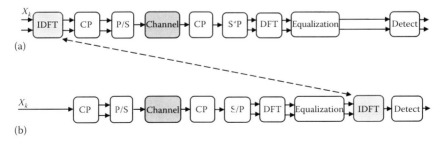

Figure 2.12   OFDM and SC-FDE systems block diagrams. (a) OFDM and (b) SC-FDE.

When comparing the two systems in Figure 2.12, it is interesting to find the similarity between them. Both use the same functional blocks, and the main difference is in the utilization of the DFT and IDFT operations. In OFDM systems, an IDFT block is placed at the transmitter to multiplex data into parallel subcarriers and a DFT block is placed at the receiver for FDE, while in SC-FDE systems, both the DFT and IDFT blocks are placed at the receiver for FDE. Thus, one can expect the two systems to have similar performance and bandwidth efficiency.

However, there are distinct differences that make the two systems perform differently. At the receiver, the OFDM system performs data detection on a per-subcarrier basis in the frequency domain, whereas the SC-FDE system does it in the time domain after the additional IDFT operation. Because of this difference, OFDM is more sensitive to a null in the channel spectrum and it requires channel coding to overcome this problem. In summary, the advantages of the SC-FDE system over the OFDM system are as follows:

- Low PAPR due to single-carrier modulation at the transmitter
- Lower sensitivity to carrier frequency offset
- Lower complexity at the transmitter, which will benefit the mobile terminal in cellular uplink communications
- A well-proven technology in several existing wireless and wireline applications

2.4 Basic Principles of the MC-CDMA System

Code division multiple access (CDMA) is a multiple access concept based on the use of wideband spread-spectrum techniques that enable the separation of signals that are coincident in time and frequency; all signals share the same spectrum. Combining OFDM and CDMA results in a multi-carrier CDMA (MC-CDMA) system [3,13,14].

2.4.1 Signal Structure

The basic MC-CDMA signal is generated by a serial concatenation of classical direct sequence (DS)-CDMA and OFDM. Each chip of the direct sequence spread data symbol is mapped onto a different subcarrier. Thus, with MC-CDMA, the chips of a spread data symbol are transmitted in parallel on different subcarriers, in contrast to a serial transmission with DS-CDMA. The number of simultaneously active users in an MC-CDMA mobile radio system is K [36].

Figure 2.13 shows the multi-carrier spectrum spreading of one complex-valued data symbol $\mathbf{d}^{(k)}$ assigned to user k. The rate of the serial data symbols is $1/T_c$. For brevity, but without loss of generality, the MC-CDMA signal generation is described for a single data symbol per user as far as possible, so that the data symbol index can be omitted. At the transmitter, the complex-valued data symbol $\mathbf{d}^{(k)}$ is multiplied by the user-specific spreading code [36]:

$$\mathbf{c}^{(k)} = (\mathbf{c}_0^{(k)}, \mathbf{c}_1^{(k)}, ..., \mathbf{c}_{N-1}^{(k)})^T \tag{2.9}$$

with length N. The chip rate of the serial spreading code $\mathbf{c}^{(k)}$ before the S/P conversion is

$$\frac{1}{T_c} = \frac{N}{T_d} \tag{2.10}$$

which is N times higher than the data symbol rate $1/T_d$. The complex-valued sequence obtained after spreading is given in vector notations by [36]

$$\mathbf{S}^{(k)} = d^{(k)}\mathbf{c}^{(k)} = (\mathbf{S}_0^{(k)}, \mathbf{S}_1^{(k)}, ..., \mathbf{S}_{N-1}^{(k)})^T \tag{2.11}$$

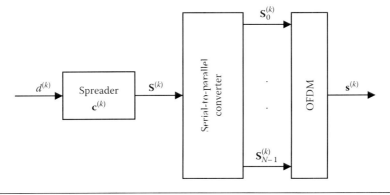

Figure 2.13 Multi-carrier spread spectrum signal generation.

A multi-carrier spread spectrum signal is obtained after modulating the components $\mathbf{S}_n^{(k)}$, $n = 0,...,N-1$ in parallel onto N subcarriers. With the multi-carrier spread spectrum, each data symbol is spread over $N-1$ subcarriers. In cases where the number of subcarriers N_c of one OFDM symbol is equal to the spreading code length N, the OFDM symbol duration with multi-carrier spread spectrum including a guard interval results in $T_s' = T_g + NT_c$. In this case, one data symbol per user is transmitted in one OFDM symbol.

2.4.2 Transmitted Signal

In the synchronous downlink, it is computationally efficient to add the spread signals of the K users before the OFDM operation as depicted in Figure 2.14. The superposition of the K sequences $\mathbf{s}^{(k)}$ results in the sequence [36]

$$\mathbf{S} = \sum_{k=0}^{K-1} \mathbf{S}^{(k)} = (\mathbf{S}_0,\mathbf{S}_1,...,\mathbf{S}_{N-1})^T \tag{2.12}$$

In matrix form, $\mathbf{S}$ can be rewritten as follows:

$$\mathbf{S} = \mathbf{C}\mathbf{d} \tag{2.13}$$

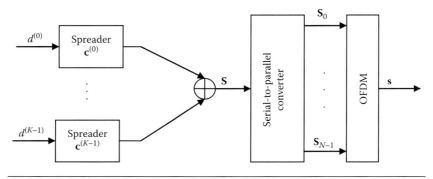

Figure 2.14   MC-CDMA transmitter.

where $\mathbf{d}$ is the vector with the transmitted data symbols of the K active users and $\mathbf{C}$ is the spreading matrix, which are given as

$$\mathbf{d} = (d^{(0)}, d^{(1)}, ..., d^{(K-1)})^T \qquad (2.14)$$

$$\mathbf{C} = (\mathbf{c}^{(0)}, \mathbf{c}^{(1)}, ..., \mathbf{c}^{(K-1)}) \qquad (2.15)$$

2.4.3 Received Signal

The received vector of the transmitted sequence $\mathbf{s}$ after OFDM demodulation and frequency de-interleaving is given by

$$R' = \mathbf{H}\mathbf{s} + \mathbf{n} = (R_C, R_1, ..., R_{N-1})^T \qquad (2.16)$$

where
  $\mathbf{H}$ is the $N \times N$ channel matrix
  $\mathbf{n}$ is the noise vector of length N

The vector R' is fed to the data detector in order to get a hard or soft estimate of the transmitted bits. The data detection needs to begin with an equalizer, which is the topic of Chapter 7.

2.4.4 Advantages of MC-CDMA

DS-CDMA is a method used to share spectrum among simultaneous users. In addition, it can exploit frequency diversity using RAKE receivers. However, in a dispersive multipath channel, DS-CDMA with a spread factor N can accommodate N simultaneous users only if highly complex interference cancellation techniques are used. This is difficult to implement in practice. MC-CDMA can handle N simultaneous users with good bit error rate (BER), using standard receiver techniques.

MC-CDMA is used to avoid excessive bit errors on subcarriers that are in a deep fade; OFDM typically applies coding. Hence, the number of subcarriers needed is larger than the number of bits or symbols transmitted simultaneously. MC-CDMA replaces this encoder by an $N \times N$ matrix operation.

MC-CDMA is likely to be the transmission method of future mobile communication devices due to its bandwidth efficiency and

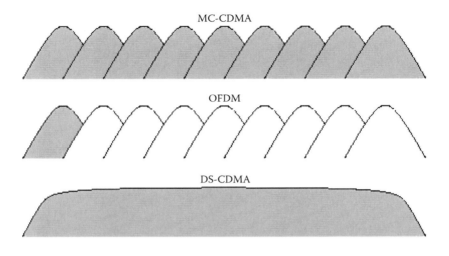

Figure 2.15 Representation of MC-CDMA, OFDM, and DS-CDMA.

inherent diversity over a fading channel. However, multi-carrier signals show highly varying envelope power waveforms, which hinder the popular employment of MC-CDMA. Our research focus is to study this phenomenon and to provide some practical solutions.

Figure 2.15 presents the MC-CDMA, OFDM, and DS-CDMA systems. It is clear that MC-CDMA and DS-CDMA use the whole bandwidth for a symbol to exploit frequency diversity, while OFDM uses a single subcarrier for it. MC-CDMA and OFDM are resilient to ISI and spectrally efficient. MC-CDMA possesses the merits of both DS-CDMA and OFDM.

2.5 PAPR Problem

The high PAPR of the transmitted OFDM signal is one of its major drawbacks. This high PAPR results in significant in-band distortion and out-of-band radiation when the signal passes through a nonlinear device such as a transmitter power amplifier (PA). The in-band distortion increases the BER and the out-of-band radiation results in an unacceptable adjacent channel interference [37]. Without the use of any PAPR reduction technique, the efficiency of power consumption at the transmitter becomes very poor.

Since OFDM signals are modulated independently in each subcarrier, the combined OFDM signals are likely to have large peak powers at certain instances. The peak power increases as the number of subcarriers increases. The peak power is generally evaluated in terms of the PAPR. The PAPR of the transmitted OFDM signal given in equation (2.1) is defined as

$$\text{PAPR} = \frac{P_{peak}}{P_{avg}} = \frac{\max\limits_{0 \le k \le K-1} \{|x(t)|^2\}}{E[|x(t)|^2]} \tag{2.17}$$

where

$E[|x(t)|^2] = \dfrac{1}{T}\displaystyle\int_0^T |x(t)|^2 \, dt$, $E[.]$ stands for the expected value

In equation (2.9), the numerator represents the maximum envelope power and the denominator represents the average power.

2.5.1 Cumulative Distribution Function of PAPR

The cumulative distribution function (CDF) of the PAPR is one of the most frequently used performance measures for PAPR reduction techniques. In the literature, the complementary CDF (CCDF) is commonly used instead of the CDF itself. The CCDF of the PAPR denotes the probability that the PAPR of a data block exceeds a given threshold. In [1], a simple approximate expression is derived for the CCDF of the PAPR of a multi-carrier signal with Nyquist rate sampling. From the central limit theorem, for a large value of K, the real and imaginary values of $x(t)$ follow a Gaussian distribution. The CDF of the amplitude z of an OFDM signal sample is given by

$$F(z) = 1 - e^{-z} \tag{2.18}$$

Next, we need to derive the CDF of the PAPR for an OFDM data block. Assuming that the signal samples are mutually independent and uncorrelated, the CDF of the PAPR for an OFDM data block can be found as

$$P(\text{PAPR} \le z) = F(z)^K = (1 - e^{-z})^K \tag{2.19}$$

The assumption that the signal samples are mutually independent and uncorrelated is not true when oversampling is applied. Also, this expression is not accurate for a small number of subcarriers, since the Gaussian assumption does not hold in this case [20,21,38,39]. It was shown that the PAPR of an oversampled signal with K subcarriers is approximated by the distribution for αK subcarriers without oversampling, where α is larger than one [2]. In other words, the effect of oversampling is approximated by adding a certain number of extra signal samples. The CDF of the PAPR of an oversampled signal is thus given by [2]

$$P(\text{PAPR} \leq z) \approx (1 - e^{-z})^{\alpha K} \tag{2.20}$$

It was found that $\alpha = 2.3$ is a good approximation for four-times oversampled OFDM signals [40]. According to equation (2.12) the probability that the PAPR exceeds PAPR_0 is given by

$$\Pr\left[\text{PAPR} > \text{PAPR}_0\right] = 1 - (1 - e^{-\text{PAPR}_0})^{\alpha K} \tag{2.21}$$

Figures 2.16 and 2.17 show the CCDF of the PAPR for an OFDM signal with different numbers of subcarriers (K = 64, 128, 256, and 512) using QPSK modulation and J = 1 and 4, respectively. It is clear from these two figures that the $\Pr$ ($\text{PAPR} > \text{PAPR}_0$) increases as the number of subcarriers increases.

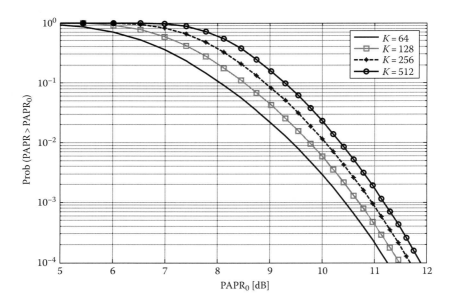

Figure 2.16 CCDF of PAPR of an OFDM signal with K = 64, 128, 256, and 512 using QPSK modulation and $J = 1$.

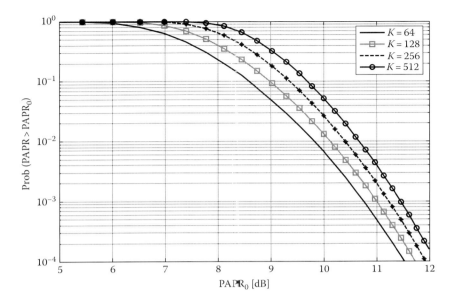

Figure 2.17 CCDF of PAPR of an OFDM signal with $K = 64$, 128, 256, and 512 using QPSK modulation and $J = 4$.

2.6 Effects of Nonlinear Power Amplifier

To determine the effect of the PAPR on OFDM system performance, nonlinear PA models must be defined. The high PAPR of OFDM requires system components with a large linear range capable of accommodating the signal dynamic range. Otherwise, nonlinear distortion occurs, which results in a loss of subcarrier orthogonality and degrades performance. Figure 2.18 shows the effects of nonlinear PA on signal spectrum (a) and (c) and signal constellation (b) and (d) using 16-QAM (a) and (b) and 64-QAM (c) and (d) with $N_{DFT} = 256$.

Ideally, the output of the PA is equal to the input times a gain factor. In reality, the PA has a limited linear region, beyond which it saturates to a maximum output level as shown in Figure 1.1.

There are two models of the PA: the solid-state power amplifier (SSPA) model and the traveling-wave tube amplifier (TWTA) model [7,41,42]. Let $S_{in}(t)$ be the input to the PA:

$$S_{in}(t) = A \exp[j\theta_i(t)] \quad (2.22)$$

The output will be

$$S_{out}(t) = G \cdot A \exp[j\{\theta_i(t) + \phi_a(t)\}] \quad (2.23)$$

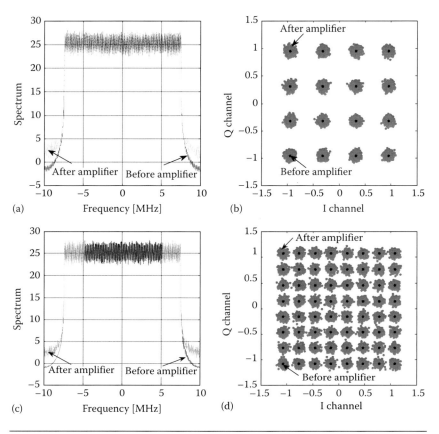

Figure 2.18   Effects of nonlinear PA on signal spectrum (a) and (c) and signal constellation (b) and (d).

where A and θ_i are the input signal amplitude and phase, respectively, G, and $\phi_a(t)$ are the amplitude and phase characteristics of the PA, respectively, which are determined according to the PA model:

$$G_{SSPA} = \frac{G_0 A}{\left[1+(A/A_{sat})^{2p}\right]^{1/2p}}, \quad \text{and} \quad \phi_a(t) = 0 \qquad (2.24)$$

$$G_{TWTA} = \frac{G_0 A}{1+(A/A_{sat})^2}, \quad \text{and} \quad \phi_a(t) = \frac{\alpha_\phi A^2}{1+\beta_\phi A^2} \qquad (2.25)$$

where

G_0 is the amplifier gain

A_{sat} is the input saturation level

P controls the sharpness of the saturation region

α_ϕ and β_ϕ are nonzero constant

Comparing equation (2.24) with equation (2.25), it is clear that the TWTA model is more nonlinear than the SSPA model [7].

The undesirable effects of PA nonlinearities can be reduced by increasing the input power backoff (IBO). For a given OFDM signal, we need to adjust the average input power so that the peaks of the signal are rarely clipped. That is, we will have to apply an IBO to the signal prior to amplification. In Figures 2.19 and 2.20, we use computer simulations to study the performance of the OFDM system using nonlinear PA with various IBO levels and $K = 64$.

Figure 2.19 gives the BER performance for the SSPA model using IBO ranging from 0 to 8 dB. At a BER of 10^{-4}, the IBO = 0 dB case suffers from a 3 dB performance loss compared to the ideal PA performance. On the other hand, the BER performance for the TWTA model given in Figure 2.20 uses IBO ranging from 0 to 16 dB, $\alpha_\phi = \pi/12$, and $\beta_\phi = 0.25$ [41]. To avoid degradation, 16 dB IBO is required; that is, 8 dB more than the SSPA case.

From this discussion, it is clear that the undesirable effects of PA nonlinearities can be reduced by increasing the IBO, which is an unsatisfactory solution, since PA efficiency reduces with IBO. The amount of IBO directly relates to both the PAPR and the PA efficiency, η_{PA};

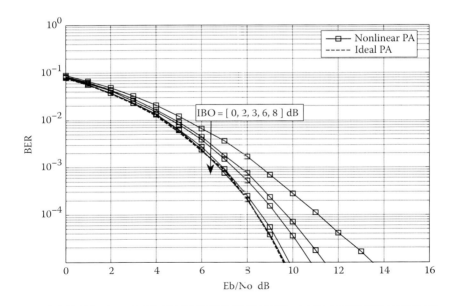

Figure 2.19 Performance of QPSK-OFDM system using SSPA with different IBO and $K = 64$.

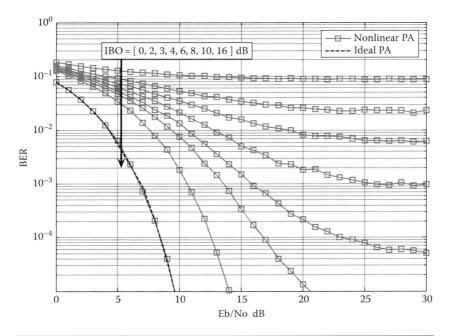

Figure 2.20   Performance of QPSK-OFDM using TWTA with different IBO and $K = 64$.

a large PAPR leads to increased IBO and reduced η_{PA}. In general, we can define IBO as the PAPR for a certain probability of clipping [33]. Consider a class A amplifier; the PA efficiency can be rewritten as [43]

$$\eta_{PA} = \frac{0.5}{\text{PAPR}} \times 100\% = \frac{0.5}{\text{IBO}} \times 100\%, \quad \text{IBO} \geq 1 \qquad (2.26)$$

The efficiency is thus inversely proportional to the IBO and the maximum efficiency, (50%) occurs at IBO = 1 (0 dB), as shown in Figure 2.21.

Therefore, increasing the IBO is not a good solution for the PAPR problem.

2.6.1 PAPR Reduction Techniques

In order to solve the PAPR problem, several schemes have been proposed. They can be grouped into three basic categories, as follows:

1. Schemes causing distortion
2. Distortionless schemes such as multiple signal representation schemes
3. Signal transformation schemes

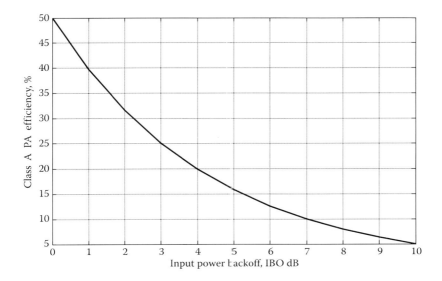

Figure 2.21 Class A power amplifier efficiency.

The schemes belonging to the first category modify the transmitted OFDM signals so that the peaks are suppressed. However, the modification operations themselves can introduce in-band noise, which may degrade BER performance. The signal clipping method [44,45] and peak windowing [46] are examples of this category.

The schemes in the second category are distortion free and the essential idea is that for one information symbol sequence, the transmitter generates multiple sequences and chooses the one with the lowest PAPR to transmit. The selective mapping (SLM) scheme and the partial transmit sequences (PTS) scheme are examples of this category [40,47–52]. These techniques improve the PAPR statistics of an OFDM signal significantly without any in-band distortion or out-of-band radiation at the cost of increasing the transmitter complexity. Also, they require side information (SI) to be transmitted to the receiver in order to make the receiver aware of what has been done at the transmitter. The errors in the detection of this side information may degrade system performance. Chapter 4 gives more details about the SLM scheme.

The third category includes schemes that are based on signal transformation before the PA. In this book, we will use continuous phase modulation (CPM) as a signal transformer. CPM is a new PAPR reduction scheme in which the high PAPR OFDM signal is transformed to a constant envelope signal (i.e., 0 dB PAPR).

The next section gives an overview of the CPM-based OFDM system. The work in this book focuses more on the second and the third categories listed here.

2.6.2 CPM-OFDM System

CPM can be used at the transmitter to transform the high PAPR OFDM signal into a low PAPR signal prior to the PA. At the receiver, the inverse transformation is performed prior to demodulation, as shown in Figure 2.22.

The phase modulator output is a constant envelope signal, which can be written as [53]

$$s(t) = Ae^{j\phi(t)} \tag{2.27}$$

where

A is the signal amplitude

$\phi(t)$ is the phase signal

The advantage of the CPM-OFDM signal, $s(t)$, is that the peak and average powers are the same, thus the PAPR is 0 dB. Figure 2.23 shows the instantaneous power of the conventional OFDM signal, $x(t)$, and the CPM-OFDM signal, $s(t)$.

According to the 0 dB PAPR of the CPM-OFDM signal, the PA can operate at the optimum (saturation) point, which has the following advantages:

- It maximizes the average transmitted power (good for coverage).
- It maximizes the PA efficiency (good for battery life).

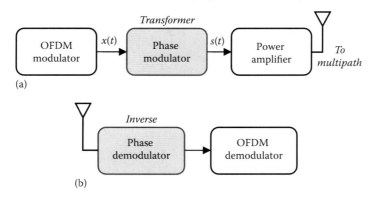

Figure 2.22   Basic concept of the CPM-OFDM system: (a) Transmitter and (b) receiver.

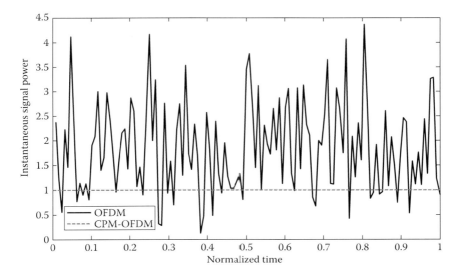

Figure 2.23   Instantaneous power of OFDM and CPM-OFDM signals.

- Since the linearity of the PA is not a must, a nonlinear PA can be used; it is generally more efficient and less expensive than the linear PA.
- CPM also exploits the frequency diversity of the channel as will be discussed in Chapters 5 and 6.

The question now is "what are the main disadvantages of CPM-based systems?" The disadvantages are as follows:

- The phase modulator transformer increases the system complexity especially at the receiver.
- The signal bandwidth grows with the modulation index of the phase modulator, which in turn reduces the bandwidth efficiency.

More details about CPM-based systems are given in Chapters 5 and 6.

3

MIMO-OFDM Space-Time Block Coding Systems

3.1 Introduction

Multiple-input-multiple-output (MIMO) communication systems have attracted much attention in recent years as a way to improve the performance of wireless mobile communications [54–56]. The MIMO technique utilizes multiple antennas at the transmitter and the receiver to improve communication link quality and/or communication capacity. MIMO wireless communication systems achieve significant capacity gains over conventional single-antenna systems. A MIMO system can provide a *spatial diversity* gain and *spatial multiplexing* gain [56]. Spatial diversity improves the reliability of communication in multipath fading channels, where it has the ability to turn multipath propagation into a benefit for users. Spatial multiplexing increases the capacity by sending multiple streams of data in parallel through multiple spatial channels.

In this chapter, we first give a general overview of MIMO concepts in Section 3.2. In Section 3.3, we describe the MIMO system model. In Section 3.4, we explain the space-time block coding (STBC) and evaluate its performance over wireless channels, and finally, in Section 3.5, we introduce the MIMO-OFDM (orthogonal frequency division multiplexing) system model

3.2 Overview of MIMO Systems

Given an arbitrary wireless communication system, we consider a link for which the transmitting and the receiving ends are equipped with multiple antennas. The idea behind MIMO techniques is that the signals on the transmit antennas at one end and the receive antennas at the other end are combined in such a way that the quality (bit error rate, BER) or the data rate (bits/s) of communication for each user will be improved.

Consider the case of two transmit antennas and one receive antenna (multiple-input-single-output, MISO), that is, there are two data paths between the transmitter and the receiver. In this scheme, there are two possibilities: either the data in the two paths are identical to each other or they are independent samples, completely different from each other. In the first case, the data seems as if it is transmitted through a single path, with the other path merely being a replica of the first one. This is a case of full correlation, in which we do not get any throughput (bits/s) advantage. However, we get a transmit diversity of two (2×1). The same is encountered in the single-input-multiple-output (SIMO) scheme, where we have a receive diversity of two (1×2).

In the second case, we deal with uncorrelated data carried by the two paths. The data streams are independent. Hence, there is no diversity, but the throughput (bits/s) is definitely higher than that in the first case. Therefore, the more the number of data paths, the higher the throughput is, provided the signals in the paths are not replicas of each other and are not correlated.

3.2.1 Spatial Diversity in MIMO Systems

Multipath fading is a significant problem in wireless communications. The basic idea behind spatial diversity techniques is to combat channel fading by having multiple copies of the transmitted signal going through independent paths. At the receiver, multiple independently faded replicas of the transmitted signal are coherently combined to achieve a more reliable reception. Let us explain how spatial diversity techniques combat channel fading. In MIMO communication systems, each pair of transmit and receive antennas provides a signal path from the transmitter to the receiver. By sending signals that carry the same information through different paths, all the paths are not expected to be in deep fade. Therefore, the loss of signal power due to fade in a single path is compensated by the signal received from a different path. Hence, the greater the diversity, the easier it is to combat fades in a channel. If the number of transmit or receive antennas tends to infinity, the diversity order tends to infinity and the channel can be approximated by an additive white Gaussian noise channel [54].

For example, in a slow Rayleigh-fading environment with one transmit and N_r receive antennas the transmitted signal passes through N_r different paths. It is well known that if the fading is independent across antenna pairs, a maximal diversity gain of N_r can be achieved. The average error probability decays with 1/SNRNr at high SNR (signal-to-noise) values, in contrast to 1/SNR for the single-antenna fading channel [57].

In a system with N_t transmit and N_r receive antennas, assuming the path gains between individual antenna pairs are independent and identically distributed (i.i.d.), the maximal diversity gain is $N_t \times N_r$. STBC [55,58] and the Alamouti transmit diversity technique [59] are some of the well-known spatial diversity techniques.

3.2.2 Spatial Multiplexing in MIMO Systems

If spatial diversity is a means to combat fading, spatial multiplexing is a way to exploit fading to increase the data throughput. Spatial multiplexing offers a linear increase in the transmission rate (or throughput) for the same bandwidth with no additional power expenses. Consider the case of two transmit and two receive antennas. The bit stream is split into two half-rate bit streams, modulated and transmitted simultaneously from both antennas. The receiver, having complete knowledge of the channel, recovers these individual bit streams and combines them to recover the original bit stream. Since the receiver has knowledge of the channel, it provides a receive diversity. However, the system has no transmit diversity, since the bit streams are completely different from each other in that they carry totally different data. Thus, spatial multiplexing increases the transmission rates proportional to the number of transmit-receive antenna pairs.

In a system with N_t transmit antennas and N_r receive antennas for high SNR, the capacity of the channel increases linearly with min (N_t, N_r) if the channel gains among the antenna pairs are i.i.d. [54,57]. Several schemes have been proposed to exploit the spatial multiplexing technique such as the Bell Labs space-time architecture (BLAST) [60].

For a given MIMO channel, both diversity and multiplexing gains can be achieved, simultaneously, but there is a fundamental trade-off between the two gains. For example, as shown in [57], the optimal diversity gain that can be achieved by any coding scheme having a multiplexing gain, m_g, is $(N_t - m_g) \times (N_r - m_g)$. This implies that out of the total resources m_g transmit and m_g receive antennas (m_g integer) are used for multiplexing and the remaining $N_t - m_g$ transmit and $N_r - m_g$ receive antennas provide the diversity. In summary, the higher spatial diversity gain comes at the price of a lower spatial multiplexing gain, and vice versa.

3.3 MIMO System Model

Consider a MIMO system with N_t transmit antennas and N_r receive antennas, as shown in Figure 3.1. The transmitted matrix is an $N_t \times 1$ column matrix x where x_i is the ith component, transmitted from antenna i. We consider the channel to be a Gaussian channel, such that the elements of x are considered to be i.i.d. Gaussian variables. If the channel is unknown at the transmitter, we assume that the transmitted signals from each antenna have equal powers of E_s/N_t with E_s representing the total transmitted power regardless of the number of antennas N_t [55].

The channel matrix H is an $N_r \times N_t$ complex matrix. The component $h_{i,j}$ of the matrix is the fading coefficient from the jth transmit

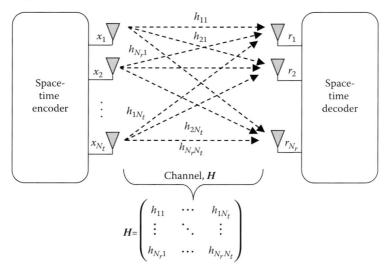

Figure 3.1   Block diagram of a MIMO system.

antenna to the ith receive antenna. For normalization purposes, we assume that the received power for each of the N_r receive antennas is equal to the total transmitted power E_s. Physically, this means that we ignore signal attenuation, antenna gains, and so on.

The received signals constitute an $N_r \times 1$ column matrix denoted by r, where each component refers to a receive antenna. Therefore, the received vector can be expressed as

$$r = Hx + n \tag{3.1}$$

Equation (3.1) can be rewritten in matrix form as follows:

$$\begin{pmatrix} r_1 \\ r_2 \\ \vdots \\ r_{N_r} \end{pmatrix} = \begin{pmatrix} b_{11} & \cdots & b_{1N_t} \\ \vdots & \ddots & \vdots \\ b_{N_r 1} & \cdots & b_{N_r N_t} \end{pmatrix} \begin{pmatrix} x_1 \\ \vdots \\ x_{N_t} \end{pmatrix} + \begin{pmatrix} n_1 \\ \vdots \\ n_{N_r} \end{pmatrix} \tag{3.2}$$

where n is the noise at the receiver, which is another column matrix of size $N_r \times 1$.

3.3.1 MIMO System Capacity

In this section, we use computer simulations to calculate MIMO system capacity. Consider a MIMO system with N_t transmit antennas and N_r receive antennas. We have two cases. In the first case we assume that the channel is unknown to the transmitter. In this case the transmitted power is equally divided among the transmit antennas [54,55]. Figure 3.2 shows the variation of the capacity with the SNR using different numbers of antennas, when the channel is unknown at the transmitter. It is clear from this figure that the capacity increases with increasing SNR and with increasing N_t and N_r.

In the second case, the channel is known to the transmitter. Under these conditions, the capacity can be increased by assigning various levels of transmitted power to various transmitting antennas using, for example, the water-filling principle [54]. In this principle, more power is allocated to the channel that is in a good condition and less power or none to the bad channels.

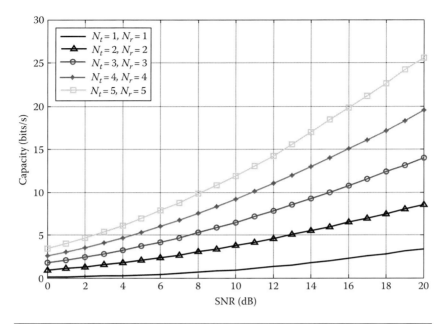

Figure 3.2 Capacity of the MIMO system for different number of antennas when the channel is unknown at the transmitter.

Figure 3.3 shows the variation of the capacity with the SNR using different numbers of antennas, when the channel is known at the transmitter with the water-filling principle. It is clear that the capacity is always higher when the channel is known than when it is unknown. This advantage reduces at high SNR values, because at high SNR values, all the channels perform equally well.

3.4 Space-Time Block Codes

In multipath environments, antenna diversity is a practical, effective, and hence a widely applied technique to reduce the effect of multipath fading. STBC is a simple transmit diversity technique in MIMO technology [54–58]. The basics of STBC were first given by the Alamouti scheme [59]. Alamouti proposed a simple scheme for a 2×2 system that achieves a full diversity gain with a simple maximum likelihood decoding algorithm. In the next section, the Alamouti scheme is explained, and we then examine the higher-order diversity schemes, whose basic approach is derived from the Alamouti scheme.

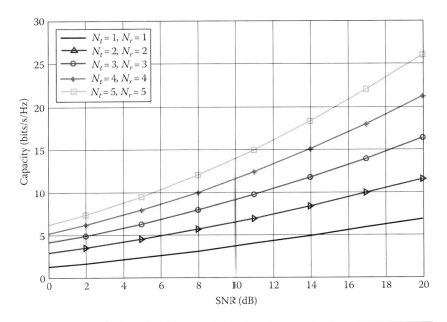

Figure 3.3 Capacity of the MIMO system for different number of antennas when the channel is known at the transmitter.

3.4.1 *Alamouti Space-Time Code*

Figure 3.4 shows the block diagram of the Alamouti space-time encoder with a 2×2 configuration. The information bits are first modulated using an M-ary modulation scheme. Then, the encoder takes a block of two modulated symbols x_1 and x_2 in each encoding operation and gives it to the transmit antennas according to the following code matrix [59]:

$$X = \begin{pmatrix} x_1 & -x_2^* \\ x_2 & x_1^* \end{pmatrix} \tag{3.3}$$

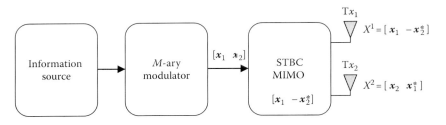

Figure 3.4 A block diagram of the Alamouti space-time encoder.

where x^* is the complex conjugate of x. The first column in X represents the first transmission period and the second column the second transmission period. The first row in X corresponds to the symbols transmitted from the first antenna and the second row corresponds to the symbols transmitted from the second antenna.

In other words, during the first symbol period, the first antenna transmits x_1 and the second antenna transmits x_2. During the second symbol period, the first antenna transmits $-x_2^*$ and the second antenna transmits x_1^*. This implies that we are transmitting both in space (across two antennas) and time (two transmission intervals). This is space-time coding.

Let us first consider the case of two transmit antennas and one receive antenna, as shown in Figure 3.5. The transmit sequence from antennas 1 and 2 is given Table 3.1.

The fading channel coefficients from transmit antennas 1 and 2 to the receive antenna are denoted by $h_1(t)$ and $h_2(t)$, respectively. Assuming that the fading coefficients are constant across two consecutive symbol transmission periods, we obtain

$$
\begin{aligned}
h_1(t) &= h_1(t+T) = h_1 = |h_1| e^{j\theta_1} \\
h_2(t) &= h_2(t+T) = h_2 = |h_2| e^{j\theta_2}
\end{aligned}
\tag{3.4}
$$

where

    T is the symbol duration

    $|h_i|$ and θ_i, $i = 1, 2$ are the amplitude gain and phase shift for the path from transmit antenna i to the receive antenna

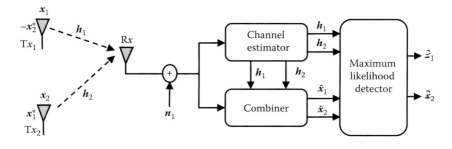

Figure 3.5   Two-antenna transmit diversity with one receiving antenna.

Table 3.1 Transmission Sequence for the Two-Antenna Transmit Diversity Scheme

	ANTENNA 1	ANTENNA 2
Time t	x_1	x_2
Time $t + T$	$-x_2^*$	x_1^*

The received signals can then be expressed as

$$\begin{pmatrix} r_1 & r_2 \end{pmatrix} = \begin{pmatrix} b_1 & b_2 \end{pmatrix} \begin{pmatrix} x_1 & -x_2^* \\ x_2 & x_1^* \end{pmatrix} + \begin{pmatrix} n_1 & n_2 \end{pmatrix} \tag{3.5}$$

Equation (3.5) can be rewritten as follows:

$$\begin{aligned} r_1 &= r(t) = b_1 x_1 + b_2 x_2 + n_1 \\ r_2 &= r(t+T) = -b_1 x_2^* + b_2 x_1^* + n_2 \end{aligned} \tag{3.6}$$

where

r_1 and r_2 are the received signals at time t and $t + T$

n_1 and n_2 are complex random variables representing receiver noise and interference at time t and $t + T$, respectively

The combiner shown in Figure 3.6 builds the following two combined signals that are sent to the maximum likelihood detector:

$$\begin{pmatrix} \tilde{x}_1 \\ \tilde{x}_2 \end{pmatrix} = \begin{pmatrix} b_1^* & b_2 \\ b_2^* & -b_1 \end{pmatrix} \begin{pmatrix} r_1 \\ r_2^* \end{pmatrix} \tag{3.7}$$

Equation (3.7) can be rewritten as follows:

$$\begin{aligned} \tilde{x}_1 &= b_1^* r_1 + b_2 r_2^* \\ \tilde{x}_2 &= b_2^* r_1 - b_1 r_2^* \end{aligned} \tag{3.8}$$

Substituting equations (3.4) and (3.6) into equation (3.8), we get

$$\begin{aligned} \tilde{x}_1 &= (\alpha_1^2 + \alpha_2^2)x_1 + b_1^* n_1 + b_2 n_2^* \\ \tilde{x}_2 &= (\alpha_1^2 + \alpha_2^2)x_2 - b_1 n_2^* + b_2^* n_1 \end{aligned} \tag{3.9}$$

where $\alpha_i = |b_i|$. It is clear from equation (3.9) that the signal $\tilde{x}_1$ depends only on x_1, while $\tilde{x}_2$ depends only on x_2. We can decide on x_1 and x_2 by applying the maximum likelihood rule on $\tilde{x}_1$ and $\tilde{x}_2$, separately. These combined signals are sent to a maximum likelihood decoder which for each transmitted symbol x_i, $i = 1, 2$, selects a symbol $\hat{x}_i$ such that $d^2(\tilde{x}_i, \hat{x}_i)$ is minimum [59], where $d^2(\tilde{x}_i, \hat{x}_i)$ is the Euclidean distance

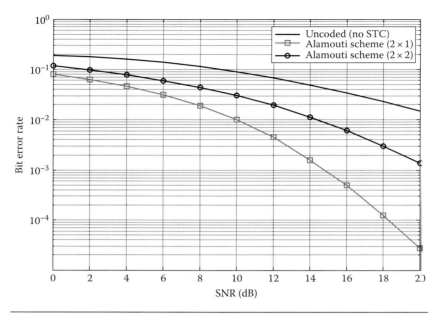

Figure 3.6 BER performance of the 16-QAM Alamouti scheme with different numbers of antennas.

between the two symbols. The complexity of the decoder is linearly proportional to the number of antennas and the transmission rate.

3.4.2 Alamouti Scheme with Multiple Receive Antennas

There are several applications where a higher-order diversity is needed. Therefore, Alamouti further extended his scheme to the case of two transmit antennas and N_r receive antennas and showed that the scheme provided a diversity order $2 \times N_r$. Let us first consider the case of two transmit and two receive antennas. The encoding and transmission sequence of the information symbols for this configuration is identical to the case of a single receiver, shown in Table 3.1. The fading channel coefficients from the first and second transmit antennas to receive antenna 1 are denoted by $h_1(t)$ and $h_2(t)$, respectively, while those to receive antenna 2 are denoted by $h_3(t)$ and $h_4(t)$, respectively, as shown in Table 3.2.

Table 3.2 Channels between the Transmit and Receive Antennas for the 2 × 2 Case

	RX ANTENNA 1	RX ANTENNA 2
Tx antenna 1	h_1	h_3
Tx antenna 2	h_2	h_4

Table 3.3   Received Signals at the Two Receive Antennas

	RX ANTENNA 1	RX ANTENNA 2
Time t	r_1	r_3
Time $t + T$	r_2	r_4

At receive antenna 1, the received signals over two consecutive symbol periods are denoted by r_1 and r_2, at times t and $t + T$, respectively, while the received signal at the receive antenna two are denoted by r_3 and r_4, at times t and $t + T$, respectively, as shown in Table 3.3.

The received signals can then be expressed as

$$\begin{pmatrix} r_1 & r_2 \\ r_3 & r_4 \end{pmatrix} = \begin{pmatrix} b_1 & b_2 \\ b_3 & b_4 \end{pmatrix} \times \begin{pmatrix} x_1 & -x_2^* \\ x_2 & x_1^* \end{pmatrix} + \begin{pmatrix} n_1 & n_2 \\ n_3 & n_4 \end{pmatrix} \tag{3.10}$$

Equation (3.10) can be rewritten as follows:

$$\begin{aligned} r_1 &= b_1 x_1 + b_2 x_2 + n_1 \\ r_2 &= -b_1 x_2^* + b_2 x_1^* + n_2 \\ r_3 &= b_3 x_1 + b_4 x_2 + n_3 \\ r_4 &= -b_3 x_2^* + b_4 x_1^* + n_4 \end{aligned} \tag{3.11}$$

where n_1, n_2, n_3, and n_4 are complex random variables representing the receiver thermal noise and interference.

The combiner can be represented by the following simple matrix form:

$$\begin{pmatrix} \tilde{x}_1 \\ \tilde{x}_2 \end{pmatrix} = \begin{pmatrix} b_1^* & b_2 & b_3^* & b_4 \\ b_2^* & -b_1 & b_4^* & -b_3 \end{pmatrix} \times \begin{pmatrix} r_1 \\ r_2^* \\ r_3 \\ r_4^* \end{pmatrix} \tag{3.12}$$

Then by applying the maximum likelihood detector, x_1 and x_2 can be estimated.

In a similar way, a $2 \times N_r$ system following the matrix form in equation (2.14) for the received signal in time interval t_1 and t_2 can be written as

$$\begin{pmatrix} r_1 & r_2 \\ r_3 & r_4 \\ \vdots & \vdots \\ r_{2m-1} & r_{2m-1} \end{pmatrix} = \begin{pmatrix} b_1 & b_2 \\ b_3 & b_4 \\ \vdots & \vdots \\ b_{2m-1} & b_{2m-1} \end{pmatrix} \times \begin{pmatrix} x_1 & -x_2^* \\ x_2 & x_1^* \end{pmatrix} + \begin{pmatrix} n_1 & n_2 \\ n_3 & n_4 \\ \vdots & \vdots \\ n_{2m-1} & n_{2m-1} \end{pmatrix} \tag{3.13}$$

The combiner can be represented by the following matrix form:

$$\begin{pmatrix} \tilde{x}_1 \\ \tilde{x}_2 \end{pmatrix} = \begin{pmatrix} h_1^* & h_2 & h_3^* & h_4 & \cdots & h_{2m-1}^* & h_{2m} \\ h_2^* & -h_1 & h_4^* & -h_3 & \cdots & h_{2m}^* & -h_{2m-1} \end{pmatrix} \times \begin{pmatrix} r_1 \\ r_2^* \\ r_3 \\ r_4^* \\ r_5 \\ r_6^* \\ \vdots \\ r_{2m-1} \\ r_{2m}^* \end{pmatrix}$$

$$(3.14)$$

Finally, in the same way, symbols x_1 and x_2 can be detected using the maximum likelihood detector.

From this discussion it is clear that the Alamouti scheme provides an overall diversity of order $2 \times N_r$. This is achieved by the key feature of the orthogonality between the sequences generated by the two transmit antennas. For this reason, this scheme was generalized to an arbitrary number of transmit antennas by applying the theory of orthogonal designs. The generalized schemes are referred to as STBC [58]. These codes can achieve the full transmit diversity of order $N_t \times N_r$, while allowing a simple maximum likelihood decoding algorithm, based only on linear processing of received signals [61].

In Figure 3.6, we use computer simulations to show the BER performance of the 16-QAM (quadrature amplitude modulation) Alamouti scheme with different antenna combinations (2×1 and 2×2). The uncoded single-input-single-output (SISO) case (no space-time coding, STC) is shown for comparison. In this simulation, it is assumed that the fading between transmit and receive antennas is mutually independent. It is clear from this figure that the Alamouti 2×1 scheme gives better performance than the SISO case. On the other hand, the 2×2 Alamouti scheme gives a better performance than the two (SISO and 2×1) cases because the order of diversity in the Alamouti scheme is high.

The BER for STBC with 8-PSK (phase shift keying) and a variable number of transmit antennas is shown in Figure 3.7. It is clear from this figure that as the number of transmit antennas increases the performance of the Alamouti scheme is improved. For example, at a BER of 10^{-3}, the 4×1 scheme gives about 5 dB gain over the 2×1 scheme.

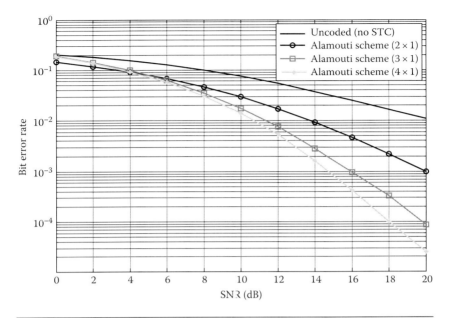

Figure 3.7   BER performance of the 8-PSK Alamouti scheme with different numbers of transmit antennas and a single receive antenna.

3.5 MIMO-OFDM Systems

For high data-rate transmission, the multipath characteristic of the environment causes the MIMO channel to be frequency-selective. As explained in Chapter 2, OFDM can transform such a frequency-selective MIMO channel into a set of parallel frequency-flat MIMO channels, and can therefore decrease the receiver complexity [62]. The combination of MIMO signal processing techniques with OFDM is regarded as a promising solution to give a good performance and support high data rate transmission in next-generation wireless communication systems. Therefore, a MIMO system employing OFDM, denoted as MIMO-OFDM, is able to achieve high bandwidth efficiency. In this section, a MIMO-STBC structure concatenated with OFDM is modeled.

3.5.1 MIMO-OFDM System Model

Figure 3.8 shows the basic model of the MIMO-OFDM system with K OFDM subcarriers, N_t transmit antennas, and N_r receive antennas.

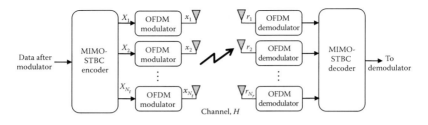

Figure 3.8 MIMO-OFDM system model.

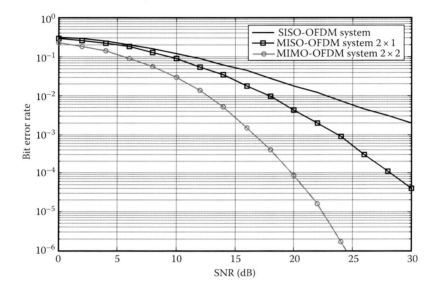

Figure 3.9 BER performance of the OFDM system with different numbers of antennas.

The received signal at antenna j from antenna i can be represented by

$$r_j(m) = \sum_{i=1}^{N_t} \sum_{l=0}^{L-1} h_{l,ji}(m)x_i(l-m) + n_j(m) \qquad (3.15)$$

where $h_{l,ji}(m)$ is the lth channel coefficient between receive antenna j and transmit antenna i. From equation (3.15), it is clear that the received signal by antenna j is the summation of transmitted signals from all transmit antennas.

Figure 3.9 shows the BER performance of the OFDM system with different numbers of transmit and receive antennas. From this figure, we can note that the performance of the 2×2 MIMO-OFDM system is better than the 2×1 MISO-OFDM system and the SISO-OFDM system.

4

PAPR Reduction Using Selective Mapping Scheme

4.1 Introduction

One of the major disadvantages of orthogonal frequency division multiplexing (OFDM) is high peak-to-average power ratio (PAPR), which may overload the power amplifier and cause in-band distortion and out-of-band radiation. The in-band distortion increases the bit error rate (BER) and the out-of-band radiation results in the unacceptable adjacent channel interference. To reduce the PAPR, several schemes have been proposed, such as clipping, coding, peak windowing, and tone reservation [40,49,63,64]. Unfortunately, most of these schemes are unable to achieve significant reduction in PAPR with low complexity and low coding overhead and without performance degradation.

The selective mapping (SLM) scheme is one of the most effective PAPR reduction schemes proposed for multiple-input-multiple-output (MIMO)-OFDM systems. It was shown that the SLM scheme can achieve several decibels of PAPR reduction and hence significantly improve the transmission power efficiency [40,48,65–68]. One of the major disadvantages of this scheme, however, is the transmission of side information (SI) bits in order to enable the receiver to recover the transmitted data blocks. These SI bits reduce the system bandwidth efficiency, as they do not contain data. Errors in the detection of these bits may lead to severe degradations in the overall system performance.

In this chapter, we propose a small-overhead SLM (s-SLM) scheme for space-time block coding (STBC)-based MIMO-OFDM systems [52]. The proposed scheme improves the system bandwidth efficiency and achieves a significantly lower BER than the individual SLM (i-SLM) and the direct SLM (d-SLM) schemes proposed in [40,69]. In addition, approximate expressions for the complementary cumulative distribution function (CCDF) of the PAPR and the average BER of the proposed s-SLM scheme are derived. The simulation results show that the proposed s-SLM scheme improves the detection probability of the SI bits, and hence gives a better performance than the i-SLM and d-SLM schemes.

In addition, the PAPR reduction performance of the proposed s-SLM scheme is improved by using an unequal power distribution approach. This approach is based on assigning powers to the different subcarriers of OFDM using an unequal power distribution strategy.

This chapter is organized as follows. In Section 4.2, the concepts of the PAPR and the CDF of the PAPR are briefly discussed. Various SLM schemes are considered in Section 4.3, where the single-input-single-output (SISO)-SLM scheme, the i-SLM scheme, the d-SLM scheme, and the proposed s-SLM scheme are explained. Section 4.4 provides a comparison between these SLM schemes. The comparison is based on the signaling overhead, complexity, and bandwidth degradation of each scheme. Section 4.5 provides an analysis of the performance of the different SLM schemes. Section 4.6 provides the numerical results and the discussion of these results. Section 4.7 presents the PAPR distribution with unequal power distribution strategy, and Section 4.8 provides concluding remarks.

4.2 PAPR of the MIMO-OFDM Signals

4.2.1 System Description and PAPR Definition

The MIMO-OFDM system adopted in this chapter depends on the Alamouti STBC scheme described in Chapter 3. This system uses K subcarriers and N_t transmitting and N_r receiving antennas. If we denote X_i as the data vector for the ith transmitting branch before the inverse fast Fourier transform (IFFT) stage for $1 \leq i \leq N_t$ and $0 \leq k < K$, then the vector x_i transmitted by the ith antenna will be the IFFT of X_i as

$$x_i = \text{IFFT}\left[X_i\right] \tag{4.1}$$

The data vector x_i is thus given by

$$x_i = \left[x_i(0), x_i(1), \ldots, x_i(k), \ldots, x_i(K-1)\right]^T \tag{4.2}$$

where $(.)^T$ denotes the matrix transpose. Because OFDM signals are modulated independently in each subcarrier, the combined OFDM signals are likely to have large peak powers at certain instances. The peak power increases as the number of subcarriers increases. As explained in Chapter 2, the peak power is generally evaluated in terms of the PAPR, which is given by

$$\text{PAPR} = \frac{\max |x_i(k)|^2}{E\left[|x_i(k)|^2\right]} \qquad (4.3)$$

where $E[.]$ is the expected value. The numerator in equation (4.3) represents the maximum envelope power and the denominator represents the average power.

The CDF of the PAPR for an OFDM data block was explained in Chapter 2. Assuming that the signal samples are mutually independent and uncorrelated, the CDF of the PAPR for an OFDM data block can be written as

$$P(\text{PAPR} \leq z) = F(z)^K = (1 - e^{-z})^K \qquad (4.4)$$

The assumption made here that the signal samples are mutually independent and uncorrelated is not true when oversampling is applied. Also, this expression is not accurate for a small number of subcarriers, since the Gaussian assumption does not hold in this case [20,21,38,39]. The effect of oversampling is approximated by adding a certain number of extra signal samples. The CDF of the PAPR of an oversampled signal is thus given by [2]

$$P(\text{PAPR} \leq z) \approx (1 - e^{-z})^{\alpha K} \qquad (4.5)$$

It was found that $\alpha = 2.3$ is a good approximation for four-times oversampled OFDM signals [40].

4.3 Selective Mapping Schemes

This section is devoted to the explanation of the existing SLM schemes and the proposed s-SLM scheme.

4.3.1 SISO-SLM Scheme

In the SISO-SLM scheme, the transmitter generates a set of sufficiently different candidate data blocks, all representing the same information as the original data block, and selects the most favorable one for transmission [52,68]. Each data block is multiplied element by element by V different phase sequences, each of length K, where, $\boldsymbol{B}^{(v)} = [b_{v,0}, b_{v,1}, ..., b_{v,K-1}]^T$, $v = 1, 2, ..., V$, resulting in V modified data blocks. Thus, the vth modified data block is $\boldsymbol{X}^{(V)} = [X_0 b_{v,0}, X_1 b_{v,1}, ..., X_{K-1} b_{v,K-1}]^T$. To include the unmodified OFDM data block in the set of modified data blocks, we set $B^{(1)}$ as the all-one vector of length K. The modified OFDM signal becomes

$$x^v(t) = \frac{1}{\sqrt{K}} \sum_{k=0}^{K-1} X_k b_{v,k} e^{j2\pi 2_k t}, \quad v = 1, 2, ..., V \qquad (4.6)$$

A block diagram of the SISO-SLM scheme is shown in Figure 4.1. Among the modified data blocks, the block $\boldsymbol{x}^{(v)}$ with the lowest PAPR is selected for transmission.

This scheme is applicable with all types of modulation and any number of subcarriers. One of its major disadvantages, however, is that it requires the transmission of SI bits about the selected phase sequence to the receiver to allow it to retrieve the original data blocks. Such a requirement has two implications. The first is the degradation in bandwidth efficiency, but this tends to be less serious when the number of subcarriers and constellation size of the employed modulation scheme are relatively large. The second and the most important issue is the BER performance, which becomes dependent on the correct estimation of the SI bits.

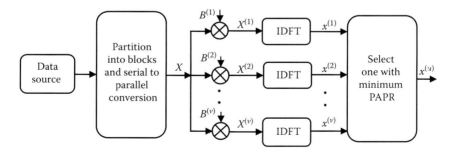

Figure 4.1   Block diagram of the SISO-SLM scheme.

For practical implementation, the SISO-SLM scheme needs V-IDFT operations, and the number of required SI bits is $N_{SI} = \lceil \log_2 V \rceil$ for each data block where $\lceil x \rceil$ denotes the smallest integer greater than or equal to x. The amount of PAPR reduction in the SLM scheme depends on the number of phase sequences V and their design. Selecting the symbol with the lowest PAPR for transmission, the probability that the PAPR exceeds a certain value of $PAPR_0$, according to (4.4), is given by

$$\Pr\left[PAPR_{SISO-SLM} > PAPR_0\right]$$

$$= \prod_{v=1}^{V} \Pr\left[PAPR > PAPR_0\right] = \left[1 - (1 - e^{-PAPR_0})^K\right]^V \quad (4.7)$$

4.3.2 MIMO-SLM Schemes

MIMO-SLM schemes are widely used in wireless communications. In this section, we discuss the traditional i-SLM and d-SLM schemes and propose an s-SLM scheme for MIMO-OFDM systems.

4.3.2.1 i-SLM Scheme In this scheme, the SISO-SLM scheme is applied to each of the N_t parallel antennas in the MIMO-OFDM system separately, as shown in Figure 4.2. For each of the parallel OFDM frames, the frame with the lowest PAPR out of the V possible ones is individually selected.

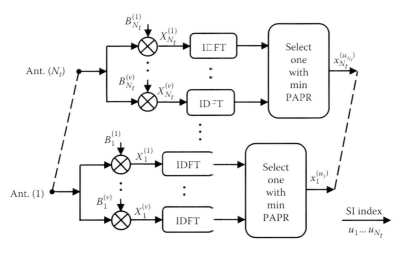

Figure 4.2 Block diagram of the i-SLM scheme with N_t transmitting antennas.

Now, $N_t \times V$ IFFT operations and hence $N_{SI} = N_t \times \lceil \log_2 V \rceil$ bits of SI are required, which means that this system is more complex than the SISO-SLM scheme and requires more redundant bits [52]. Selecting the set with the lowest average PAPR for transmission, the probability that the lowest PAPR over the N_t transmitting antennas exceeds PAPR_0 is given by

$$\Pr\left[\mathrm{PAPR}_{\mathrm{MIMO\ i\text{-}SLM}} > \mathrm{PAPR}_0\right] = 1 - \left\{1 - \left[1 - (1 - e^{-\mathrm{PAPR}_0})^K\right]^V\right\}^{N_t}$$

(4.8)

4.3.2.2 d–SLM Scheme   In [69], the authors introduce a new version of the SLM technique for PAPR reduction called the d-SLM. The main advantage of the d-SLM scheme over the i-SLM scheme is its ability to use the advantages of MIMO transmission, which means that the PAPR performance improves as the number of transmitting antennas increases. This is achieved at the expense of increasing the number of SI bits (redundancy), and so a further bandwidth loss takes place. As shown in [69], the number of SI bits required for the d-SLM scheme is $N_{SI} = N_t \times \lceil \log_2(N_t(V-1) + 1) \rceil$ with the same complexity of $N_t \times V$ IFFT operations as in the i-SLM scheme. In the d-SLM scheme, the probability that the lowest PAPR over the N_t transmitting antennas exceeds PAPR_0 is given by [70]

$$\Pr\left[\mathrm{PAPR}_{\mathrm{d\text{-}SLM}} > \mathrm{PAPR}_0\right] = \left[1 - \left(1 - e^{-\mathrm{PAPR}_0/\Delta}\right)^K\right]^{N_t V}$$

(4.9)

where Δ is the PAPR loss compared to the SISO-SLM scheme with $N_t V$ candidates, $\Delta = 0.45$ dB for $V = 4$ [70].

4.3.2.3 Proposed s–SLM Scheme   This section presents the proposed s-SLM scheme for STBC-based MIMO-OFDM systems [52]. The main advantage of the proposed scheme is the reduction of the signaling overhead as compared to the i-SLM and the d-SLM schemes. Also, the detection probability of the SI bits in the proposed scheme improves because of the space-time diversity (STD) achieved by the different transmitting and receiving antennas, and hence the overall BER improves as compared to the i-SLM and the d-SLM schemes.

In the proposed scheme, the N_t different subcarrier vectors X_i are multiplied subcarrier-wise by the same phase sequence, which is one

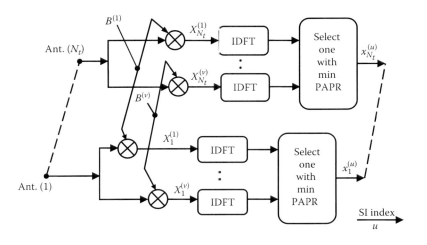

Figure 4.3 Block diagram of the s-SLM scheme with N_t transmitting antennas.

of the V sequences, resulting in V sets composed of N_t different sub-carrier vectors as shown in Figure 4.3. For each set, the N_t different subcarrier vectors are transformed by N_t separate K-points IFFT operations.

The set with the lowest average PAPR over all transmitting antennas is chosen. So only $N_{SI} = \lceil \log_2 V \rceil$ bits of SI are required [52]. However, no complexity reduction is achieved as we still need $N_t \times V$ IFFT operations. Selecting the set with the lowest average PAPR for transmission, the probability that the lowest PAPR over all the N_t transmitting antennas exceeds PAPR_0 is given by

$$\Pr\left[\text{PAPR}_{\text{MIMO-sSLM}} > \text{PAPR}_0\right] = \left(1 - (1 - e^{-\text{PAPR}_0})^{N_t \cdot K}\right)^V \quad (4.10)$$

Since the subcarrier vectors with the lowest average PAPR are constructed using the same mapping vectors in all transmitting antennas, the N_t transmitting antennas can bear the same SI bits. The STBC diversity for the SI bits is guaranteed by the increase of N_t. To obtain full frequency diversity during the transmission of OFDM symbols, the same SI bits can be inserted into an appropriate subcarrier to guarantee the largest possible frequency distance over the N_t transmitting antennas. When the MIMO-OFDM signal uses an M-QAM constellation, each subcarrier can carry $\log_2 M$ SI bits.

4.4 Comparison between Different SLM Schemes

Two factors are considered in the comparison between the different SLM schemes: the number of SI bits and bandwidth degradation.

4.4.1 Number of SI Bits

This section provides a comparison between the different SLM schemes considering the required number of SI bits. Table 4.1 compares the number of SI bits (redundancy) and the number of IFFT operations (complexity) required for the SLM schemes with $N_t = 4$. It is clear from the table that the s-SLM scheme has the smallest number of SI bits, when compared to the other MIMO-SLM schemes (i-SLM and d-SLM).

4.4.2 Bandwidth Degradation

Transmission of SI bits leads to further bandwidth loss and reduction in the energy per bit in the information symbols. Such bandwidth loss is particularly significant when modulation schemes with small constellation sizes, such as binary and quaternary phase shift keying (BPSK and QPSK), are used. The percentage of bandwidth degradation (B_{loss}) due to the SI bits can be calculated as follows [71]:

$$B_{loss} = \left(1 - \frac{mK - (N_{SI}/mR_c)}{mK}\right) \times 100 = \frac{N_{SI}}{m^2 KR_c} \times 100 \quad (4.11)$$

where
   m is the number of bits per symbol
   N_{SI} is the number of SI bits
   R_c is the coding rate ($R_c \leq 1$, $R_c = 1$ refers to the uncoded system)

Table 4.1   Number of SI Bits and IFFT Operations Required for the Different SLM Schemes with $N_t = 4$

	SISO-SLM		MIMO i-SLM		MIMO d-SLM		MIMO s-SLM	
V	N_{SI}	NO. OF IFFT OPERATIONS	N_{SI}	NO. OF IFFT OPERATIONS	N_{SI}	NO. OF IFFT OPERATIONS	N_{SI}	NO. OF IFFT OPERATIONS
4	2	4	8	16	16	16	2	16
8	3	8	12	32	20	32	3	32
16	4	16	16	64	24	64	4	64

Table 4.2 B_{loss} for the Different SLM Schemes for Uncoded and Coded OFDM

	SISO-SLM		MIMO i-SLM		MIMO d-SLM		MIMO s-SLM	
V	UNCODED	CODED	UNCODED	CODED	UNCODED	CODED	UNCODED	CODED
4	1.563	3.125	6.250	12.5	12.5	25	1.563	3.125
8	2.344	4.688	9.375	18.75	15.625	31.25	2.344	4.688
16	3.125	6.250	12.5	25	18.75	37.5	3.125	6.250

Table 4.2 gives the value of B_{loss} in the case of both channel-coded ($R_c = 1/2$) and uncoded BPSK for $K = 128$ and $N_t = 4$. It is clear from the table that the s-SLM scheme achieves the smallest bandwidth loss, when compared to the i-SLM and d-SLM schemes. In addition, it improves the detection probability of SI bits and the overall BER when compared with the SISO-SLM scheme, as explained in the next section.

4.5 Performance Analysis of Different SLM Schemes

The correct detection of the SI bits is essential for the success of all SLM schemes as any incorrect decision at this stage would lead to incorrect demodulation of the whole block of data symbols and a significant degradation in system performance. That is, the system BER is dependent on the correct detection of the SI bits, and this gives an advantage to the proposed scheme, where the detection probability of the SI bits improves due to the STD diversity resulting from increasing the number of transmitting and receiving antennas.

At the receiving end, the receiver must be able to separate the transmitted OFDM signal from the embedded SI bits. In the proposed s-SLM scheme, it is possible that only one transmitting antenna bears the SI bits and the N_r receiving antennas receive these bits. The detection probability of the SI bits can be obtained as follows [40]:

$$P_d = 1 - N_{SI} P_b \qquad (4.12)$$

with

$$P_b = \left(\frac{1-\lambda}{2}\right)^D \sum_{n=0}^{D-1} \binom{D-1+n}{n}\left(\frac{1+\lambda}{2}\right)^n \qquad (4.13)$$

where $\lambda = [\gamma_o/(1 + \gamma_o)]^{0.5}$; here γ_o is the average SNR per antenna and D represents the number of diversity channels carrying the same information-bearing signal [48]. Since the s-SLM scheme uses the same mapping vector in all transmitting antennas, the STBC gain and the frequency diversity gain can be exploited. Hence the possible diversity can be approximated by $D = N_t \times N_r$ with the help of the STD diversity in the s-SLM scheme. We can observe that in the proposed scheme, the performance gets improved as the number of receiving antennas increases. With the increase of N_t, however, the gain of the frequency diversity may be smaller due to the decrease of the possible frequency distance.

If we assume that the $\boldsymbol{B}^{(v)}$ vectors for $1 \leq v \leq V$ are statistically independent and that the probability of false detection of the SI bits is $1 - P_d$, the overall BER for all SLM schemes denoted by P_0 can be approximated by [48,52]

$$P_0 \approx P_e \cdot P_d + 1 - P_d \tag{4.14}$$

where P_e is the BER of the MIMO-OFDM system given that the SI bits are correctly detected. It can be expressed in a form similar to equation (4.13). In equation (4.14), the false detection probability of the SI bits $(1 - P_d)$ is very small as a result of the STD diversity.

4.6 Results and Discussion

In this section, both theoretical expressions and simulations are used to test the proposed MIMO s-SLM scheme and compare it with the SISO-SLM, MIMO i-SLM, and d-SLM schemes. Figures 4.4 and 4.5 show a comparison in the PAPR performance between the different SLM schemes using BPSK modulation with $K = 128$ and $\alpha = 1$ (nonoversampling case), for $N_t = 2$ and 4, respectively. For $\Pr[\text{PAPR} > \text{PAPR}_0] = 10^{-3}$, in the case of $N_t = 2$, the PAPR performance degradation of the proposed s-SLM scheme when compared with the i-SLM scheme is about 0.3 dB at $V = 4$. This degradation is still smaller than that in References [40] and [48] where the performance degradation is ≥0.5 dB.

The performance degradation in the proposed scheme becomes 0.75 dB compared to that of the d-SLM scheme. We can also see that as the number of transmitting antennas increases, the PAPR reduction performance in the i-SLM and s-SLM schemes is degraded regardless of V, as shown in Figure 4.5. On the other hand, the PAPR reduction performance of the d-SLM scheme improves as the number

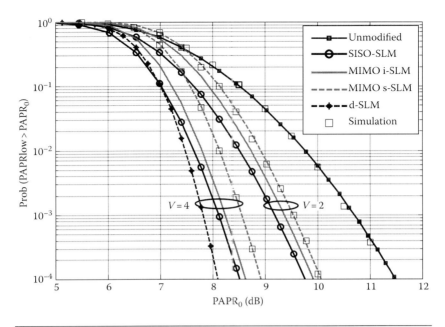

Figure 4.4 PAPR performance of the different S_M schemes for $K = 128$ and $N_t = 2$.

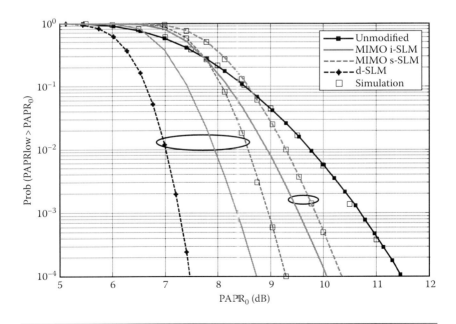

Figure 4.5 PAPR performance of the different S_M schemes for $K = 128$ and $N_t = 4$.

of transmitting antennas increases. According to the obtained results, the performance degradation in the proposed scheme is small as compared to the advantages we gained—improved system bandwidth efficiency, detection probability of SI bits, and finally the overall BER, as shown Figures 4.6 and 4.7.

In the next section, the PAPR reduction performance of the proposed s-SLM scheme is improved by using an unequal power distribution strategy.

Figure 4.6 shows the detection probability of the SI bits using equation (4.12) for the different SLM schemes with $K = 128$, $V = 4$, and $N_t = N_r = 2$. The obtained results are also verified by simulations. These results show that the detection probability of the SI bits in the proposed s-SLM scheme improves when compared to the other SLM schemes, due to the STD diversity resulting from increasing the number of transmitting and receiving antennas. For example, at an SNR = 6 dB, the detection probability of the proposed s-SLM is 97% compared to 87%, 75%, and 79% for the i-SLM, d-SLM, and the SISO-SLM schemes, respectively.

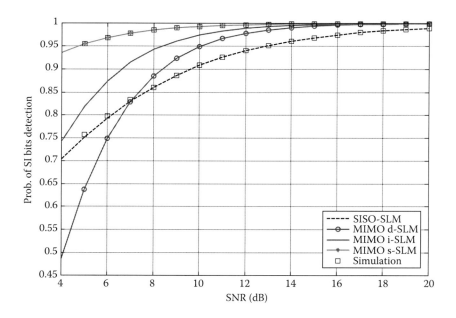

Figure 4.6 Probability of SI bits detection for the different SLM schemes with $K = 128$, $V = 4$, and $N_t = N_r = 2$.

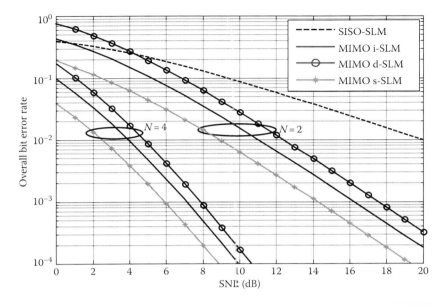

Figure 4.7   Overall BER for the different SLM schemes with $K = 128$, $V = 4$, and $N_t = 2$.

Figure 4.7 shows the overall BER in the case of erroneous SI bits using equation (4.14) for the different SLM schemes with $K = 128$ and $V = 4$. The SNRs required in the i-SLM and d-SLM schemes with $N_r = 2$ are 2 and 3.3 dB higher than that required in the proposed s-SLM scheme, respectively, to achieve a BER of 10^{-3}. This advantage of the proposed scheme is attributed to the STD diversity for detecting the SI bits. On the other hand, the proposed s-SLM scheme provides a BER $= 9.3 \times 10^{-4}$ at an SNR $= 6$ dB compared to 2.4×10^{-3}, 4×10^{-3}, and 0.19 for the i-SLM, d-SLM, and SISO-SLM schemes using $N_r = 4$, respectively.

4.7 PAPR with Unequal Power Distribution

In this section, the PAPR reduction performance of the proposed s-SLM scheme is improved by using an unequal power distribution strategy. The PAPR distributions presented in equation (4.7) and in [47–52,68] are obtained under the assumption that all subcarriers are active and allocated equal power. This assumption may not be valid due to the following facts. First, in all realistic OFDM systems, usually only a subset of subcarriers is used to carry information (active subcarriers) and the other subcarriers (inactive subcarriers) are set to zero. Second, due to efficiency considerations, transmission power should be allocated to active subcarriers. Third, power allocation may vary depending on the different constellations used by the different active subcarriers and their SNRs.

The proposed approach is based on assigning powers to the different subcarriers of OFDM using an unequal power distribution strategy (assigning powers to active subcarriers, only). The OFDM symbol structure of the new approach consists of three types of subcarriers as shown in Figure 4.8: data subcarriers for data transmission, pilot subcarriers for channel estimation and synchronization, and null subcarriers for guard bands. The data and pilot subcarriers represent the active subcarriers with K_{active} denoting their number. The null subcarriers are named as inactive subcarriers with $K_{inactive}$ denoting their number. If the subcarrier at DC(0) is nonzero, it is active; otherwise it is inactive.

Table 4.3 shows the number of active subcarriers in some popular standards. In this table, K_0 represents the subcarrier at DC, I_s represents inactive subcarriers at DC, and A_s represents active subcarriers at DC [7,72]. For example, in the IEEE 802.11a standard, out of all 64 subcarriers, 52 subcarriers are active (data and pilots subcarriers) and 12 are inactive.

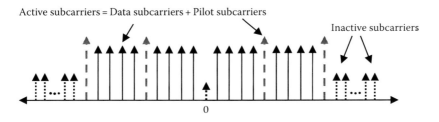

Figure 4.8 The OFDM symbol structure with unequal power distribution strategy.

Table 4.3 Number of Active and Inactive Subcarriers in Some Popular Standards with K_0 Representing the Subcarrier at DC

	802.11a	DAB I	DAB II	DVB I	DVB II	DVB III
K	64	256	512	2048	2048	4096
K_{active}	52	192	384	1536	1705	3409
K_0	I_s	I_s	I_s	I_s	A_s	A_s

In this case, the equivalent complex baseband OFDM signal can be rewritten as

$$x(t) = \frac{1}{\sqrt{K}} \sum_{k=-N}^{N} X_k e^{j2\pi f_k t} \tag{4.15}$$

where $K = K_{active} + K_{inactive}$. In equation (4.15), $N = K_{active}/2$ if the subcarrier at DC is inactive; otherwise $N = (K_{active} - 1)/2$.

Let P_k be the power allocated to the kth subcarrier, we assume that equal powers are allocated to active subcarriers, and thus P_k is constant. Therefore, when all active subcarriers are allocated equal power, equation (4.7) can be rewritten as

$$\Pr\left[PAPR_{conv.-SLM} > PAPR_0\right] = \left[1 - (1 - e^{-PAPR_0})^{2N}\right]^V \tag{4.16}$$

Note that in equation (4.16), we assume that the subcarrier at DC is inactive.

Figure 4.9 shows the CCDF of the PAPR for the WLAN 802.11a and the DAB I standards according to Table 4.3, with BPSK and equal power allocated to each active subcarrier. According to this figure, we notice that, at $\text{Prob}(PAPR > PAPR_0) = 10^{-3}$, the PAPR

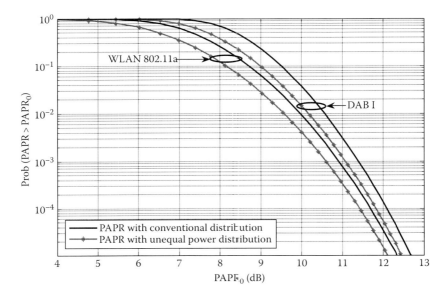

Figure 4.9 Comparison between the different CCDFs of the PAPR in the WLAN 802.11a and the DAB I standards.

reduction performance with unequal power distribution is better than that of conventional distribution (all subcarriers are active and allocated equal power) by about 0.4 dB.

Figure 4.10 shows the performance of the conventional SISO-SLM scheme with an unequal power distribution strategy. Usually, K_{active} is about $3K/4$ and $N = K_{active}/2$ if the subcarrier at DC is inactive, that is, $N \approx 3K/8$ [7]. Note that in this figure, system 1 represents the conventional SISO-SLM scheme, in which the PAPR distributions are obtained under the assumption that all subcarriers are active and allocated equal power (conventional distribution) as in equation (4.7). System 2 represents the SISO-SLM scheme with the unequal power distribution strategy.

The obtained results in Figure 4.10 show that the new approach with the unequal power distribution strategy improves the PAPR reduction performance. Further improvements can be made by increasing the number of phase sequence candidates, V. For example, at $\text{Prob}(\text{PAPR} > \text{PAPR}_0) = 10^{-3}$ and $V = 8$, the SISO-SLM system 2 scheme can achieve about 0.45 dB more PAPR reduction as compared with the SISO-SLM system 1 scheme.

This new approach can be applied to the proposed MIMO s-SLM scheme to improve its PAPR reduction performance. Figure 4.11 shows

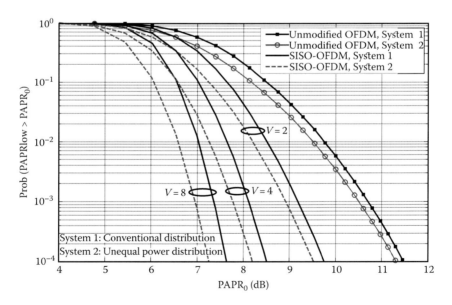

Figure 4.10 Performance of the conventional SISO-SLM scheme with conventional distribution and unequal power distribution strategies.

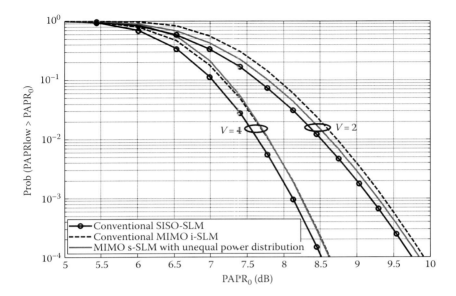

Figure 4.11   PAPR performance of the proposed MIMO s-SLM scheme with unequal power distribution strategies.

the PAPR performance of the proposed MIMO s-SLM scheme with unequal power distribution strategy. It is clear from this figure that the proposed MIMO s-SLM scheme with the unequal power distribution strategy provides better PAPR reduction than the MIMO i-SLM scheme at $V = 2$, and nearly the same performance at $V = 4$.

In the following section, we propose a reduced complexity selective mapping (RC-SLM) scheme. The proposed scheme is based on partitioning the frequency domain symbol sequence into several sub-blocks, and then each sub-block is multiplied by different phase sequences whose length is shorter than those used in the conventional SLM scheme. Then we use a kind of low complexity conversion to replace the IFFT blocks. The performance of the proposed RC-SLM scheme along with the unequal power distribution strategy is studied with computer simulation.

4.8 Proposed RC-SLM Scheme

It is clear from Figure 4.1 that the conventional SLM scheme requires a large amount of IFFT calculation and multiplication with phase sequences which is proportional to the number and the length of the phase sequences. To reduce the complexity of the conventional SLM scheme, the authors in [73] developed conversion matrices (originated or modified from some phase rotation vectors of period four) to replace more IFFT blocks in the conventional SLM scheme. For an OFDM system with K subcarriers and J times oversampling, these conversion matrices can replace some of the IFFT blocks in the conventional SLM method, where each conversion process requires only $3JK$ complex additions to compute a JK-point IFFT. As compared to the conventional SLM scheme, the approach in [73] reduces the computational complexity but with performance degradation in terms of PAPR reduction and BER performance.

In our proposed RC-SLM scheme [74], the computational complexity reduction is based on reducing both the length of the phase sequences and the number of IFFT calculations. The length of the phase sequences can be reduced by partitioning the frequency domain symbol sequence X into L sub-blocks of length $M = K/L$. Then each sub-block is multiplied by different phase sequences, $p^{(m)}$, whose length is M which is shorter than those used in the conventional SLM scheme. As a result, compared with the approaches proposed in [73,75], the K-point IFFT is replaced with M-point IFFT and the computational complexity reduced considerably.

Then we use a kind of low-complexity conversion such as that in [73] or [75] to replace the IFFT blocks. Each conversion process requires only $3JK$ complex additions to compute a JK-point IFFT. With these conversion matrices, we may use only one IFFT output to generate other candidate signals. Finally, the PAPR is calculated for all the sub-block combinations and the sub-block combination whose sum exhibits the lowest PAPR is selected for transmission as in the conventional SLM scheme. A block diagram of the proposed RC-SLM scheme is shown in Figure 4.12 [74].

The low complexity conversion used in Figure 4.12 is named *replacement conversion* (R); this conversion is used to replace the IFFT block in the conventional SLM scheme. The idea of this conversion is explained in Figure 4.13, the signals x_1 and x_2 are the IFFT of X_1 and X_2, respectively, with $X_1 = X \cdot p^{(1)}$ and $X_2 = X \cdot p^{(2)}$ and $p^{(m)}$ is the phase

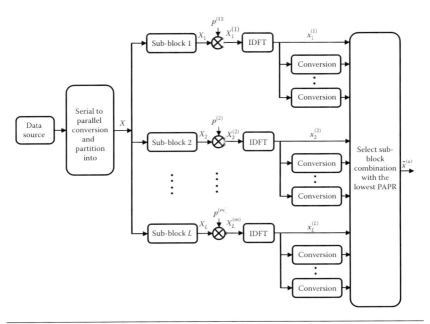

Figure 4.12   Block diagram of the proposed RC-SLM scheme.

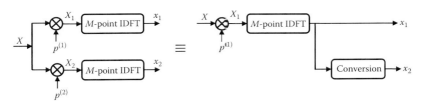

Figure 4.13   The conversion replacement for IFFT computation.

rotation vector whose length is M. In this conversion, **R**, we use one IFFT output, x_1, to get another IFFT output, x_2, as follows [74]:

$$x_1 = \text{IFFT}\{X_1\} = WX_1 = WXp^{(1)} \tag{4.17}$$

$$x_2 = \text{IFFT}\{X_2\} = WX_2 = WXp^{(2)} \tag{4.18}$$

where W represents the IFFT matrix as follows:

$$W = \frac{1}{M}\begin{bmatrix} 1 & 1 & 1 & \cdots & 1 \\ 1 & W_M^1 & W_M^2 & \cdots & W_M^{M-1} \\ 1 & W_M^2 & W_M^4 & \cdots & W_M^{2(M-1)} \\ \vdots & \vdots & \vdots & \ddots & \vdots \\ 1 & W_M^{M-1} & W_M^{2(M-1)} & \cdots & W_M^{(M-1)(M-1)} \end{bmatrix} \tag{4.19}$$

In equation (4.19), $W_M = \exp\{j2\pi/M\}$. As in Section 2.2, we usually use $p^{(1)}$ as the all-one vector of length M to include an unmodified OFDM data sub-block in the set of the modified data sub-blocks. In equation (4.17), substitute for $p^{(1)}$ with all one vector results in

$$x_1 = \text{IFFT}\{X_1\} = WX_1 = WX \tag{4.20}$$

From equation (4.20), we can write

$$X = W^{-1}x_1 \tag{4.21}$$

where W^{-1} represents the FFT matrix. Using equation (4.21), equation (4.18) can be rewritten as

$$x_2 = WW^{-1}x_1 p^{(2)} = I_M x_1 p^{(2)} \tag{4.22}$$

where I_M is the identity matrix. The above discussion means that the next candidate can be obtained simply by using the conversion **R** which is simply equivalent to multiplying the signal x_1 with the identity matrix, I_M, and then by the next phase sequence. The nth candidate can be rewritten as

$$x_n = WW^{-1}x_1 \, p^{(n)} = I_M x_1 p^{(n)} \tag{4.23}$$

4.8.1 Computational Complexity Comparison

It is well known from the literature that a JK-point IFFT requires $(JK/2 \log_2 JK)$ complex multiplications and $(JK \log_2 JK)$ complex additions. Therefore, the conventional SLM, (C-SLM) scheme with V individual phase sequences requires V IFFT operations. This means that the total number of multiplication N_m and addition N_a in the conventional SLM scheme can be expressed as [74,76]

$$N_{m,\text{C-SLM}} = V \times \frac{JK}{2} \log_2 JK + V \times JK \tag{4.24}$$

$$N_{a,\text{C-SLM}} = V \times JK \log_2 JK \tag{4.25}$$

The second term in equation (4.24) represents the amount of multiplication of data vector X with V phase sequences before the IFFT stages.

On the other hand, in the case of the proposed RC-SLM scheme, as shown in Figure 4.12, there are L sub-blocks each requiring one multiplication to get the phase-modulated data before the IFFT stage and one IFFT stage per sub-block. The conversion $\mathbf{R}$ requires $3JM$ complex additions and zero multiplications [73]. For V phase sequences, there are $V-1$ number of conversion $\mathbf{R}$ required per sub-block. According to this discussion, the total number of multiplication N_m and addition N_a in the proposed RC-SLM scheme can be expressed as [74]

$$N_{m,\text{RC-SLM}} = L \times \frac{JM}{2} \log_2 JM + L \times JM \qquad (4.26)$$

$$N_{a,\text{RC-SLM}} = L \times JM \log_2 JM + 3L(V-1)JM \qquad (4.27)$$

Let us now compute the computational complexity reduction ratio (CCRR) of the proposed RC-SLM scheme, over the conventional C-SLM scheme, which is defined as

$$\text{CCRR} = \left(1 - \frac{\text{The proposed RC-SLM scheme complexity}}{\text{The conventional C-SLM scheme complexity}}\right) \times 100\%$$
$$(4.28)$$

Figure 4.14 shows a comparison between the proposed RC-SLM scheme and the conventional SLM scheme in terms of the total number of complex operations. The parameters used in this figure are as follows: $J = 4$, $V = 4$, and $L = 2$. It is clear from this figure that the proposed RC-SLM scheme has much lower complexity than the conventional SLM. For example, at $K = 512$, the proposed RC-SLM scheme achieves 77.3% and 53% reduction of the number of required multiplications and additions, respectively, when compared with the conventional SLM scheme.

Figure 4.15 shows a comparison between the proposed RC-SLM scheme and the schemes proposed in References [73] and [76] in terms of total number of complex operations. It is clear from this figure that the proposed RC-SLM scheme has the lowest computational complexity when compared to the other schemes.

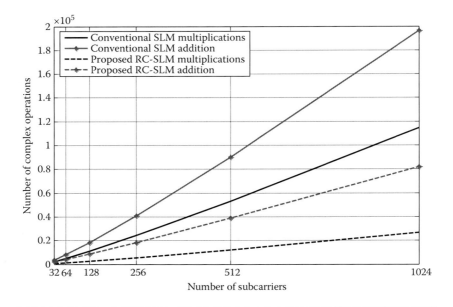

Figure 4.14 Comparison between the proposed RC-SLM scheme and the conventional SLM scheme in terms of total number of complex operations.

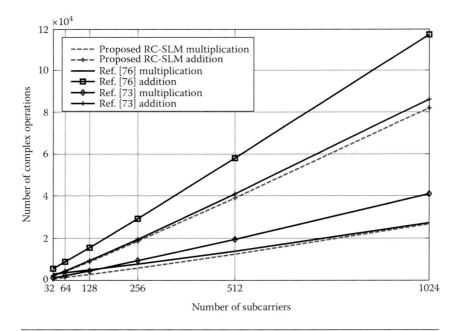

Figure 4.15 Comparison between the proposed RC-SLM scheme and the schemes proposed in References [73] and [76] in terms of total number of complex operations.

Table 4.4 CCRR of the Proposed RC-SLM Scheme Compared to the C-SLM Scheme in Terms of V

	NUMBER OF MULTIPLICATION, N_m			NUMBER OF ADDITION, N_a		
V	C-SLM	RC-SLM	CCRR (%)	C-SLM	RC-SLM	CCRR (%)
2	5,632	2,560	54.6	9,216	5,632	39
4	11,264	2,560	77.3	18,432	8,704	52.8
8	22,528	2,560	88.9	36,864	14,848	60

Table 4.5 CCRR of the Proposed RC-SLM Scheme Compared to the C-SLM Scheme in Terms of L

	NUMBER OF MULTIPLICATION, N_m			NUMBER OF ADDITION, N_a		
L	C-SLM	RC-SLM	CCRR (%)	C-SLM	RC-SLM	CCRR (%)
2	11,264	2,560	77.3	18,432	8,704	53
4	11,264	5,120	54.5	18,432	17,408	6

The CCRR of the proposed RC-SLM scheme over the conventional C-SLM scheme for different values of V is given in Table 4.1 [74], using $K = 128$, $J = 4$, $L = 2$, and $M = 64$. Here, we assume that the two schemes generate the same numbers of candidates. It is shown in Table 4.4 that the number of multiplications in the proposed RC-SLM scheme does not depend on the number of phase sequences, V. Moreover, the computational complexity of the RC-SLM scheme is reduced rapidly with the increase of V.

Table 4.5 shows the CCRR of the proposed RC-SLM scheme over the conventional C-SLM scheme for different values of L using $K = 128$, $J = 4$, and $V = 4$. It is clear from this table that the computational complexity of the proposed RC-SLM scheme is still lower than that of the conventional C-SLM scheme even for a larger value of L.

4.8.2 Simulation Results and Discussion

In this section, the PAPR reduction performances of the conventional SLM and the proposed RC-SLM schemes are investigated theoretically and by computer simulations. The OFDM system used for simulations has $K = 128$ subcarriers, QPSK modulation, and 20 MHz bandwidth. The performance is evaluated in terms of the complementary cumulative distribution function (CCDF) and by using an oversampling factor of $J = 4$. The phase rotation vectors used in the simulation are randomly generated, that is, their elements are randomly selected from the set $\{\pm 1, \pm j\}$. Here, we assume that the two schemes use the same number of candidates. The number of phase sequence candidates, V, and the number of sub-blocks, L, are changed during the simulation.

Figure 4.16 shows the performance of the conventional C-SLM scheme with the unequal power distribution strategy explained in Section 2.3. Usually, K_{active} is about $3K/4$, $N = K_{active}/2$ if the subcarrier at DC is inactive, that is, $N \approx 3K/8$ [7]. Note that in this figure, system 1 represents the C-SLM scheme in which the PAPR distributions are obtained under the assumption that all subcarriers are active and allocated equal power as in equation (4.14) and in References [52,67,68,77,78], while system 2 represents the C-SLM scheme with the unequal power distribution strategy [74].

The obtained results in Figure 4.16 show that the new approach with unequal power distribution strategy can improve the PAPR reduction performances. Further improvements can be obtained by increasing the number of phase sequence candidates, V. For example, at $\text{Prob}(\text{PAPR} > \text{PAPR}_0) = 10^{-3}$ and $V = 8$, the C-SLM system 2

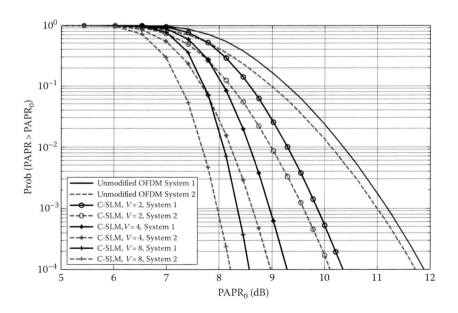

Figure 4.16   Performance of the conventional C-SLM scheme with equal and unequal power distribution strategies.

scheme can achieve about 0.5 dB more PAPR reduction when compared with the C-SLM system 1 scheme.

Figure 4.17 shows the PAPR performance of the proposed RC-SLM scheme with various (L, V) combinations. The unmodified OFDM system and the conventional SLM scheme are used for comparison. It is clear that the proposed RC-SLM scheme with $L = 2$ and $V = 4$ (i.e., with more candidates) achieves better performance than the one that uses $L = 4$ and $V = 2$. It also outperforms the conventional C-SLM scheme. In the following results we will use $L = 2$ and $V = 4$.

Figure 4.18 shows a performance comparison between the proposed RC-SLM scheme with the unequal power distribution strategy using $L = 2$ and the conventional C-SLM scheme. It is clear from this figure that the proposed RC-SLM scheme outperforms the conventional C-SLM scheme in terms of PAPR reduction performance. For example, at $\text{Prob}(\text{PAPR} > \text{PAPR}_0) = 10^{-3}$ and $V = 8$, the proposed RC-SLM scheme provides PAPR performance improvements of about 0.3 dB over the conventional C-SLM scheme, unlike the schemes presented in References [73,75,79], which achieve low computational complexity at the cost of PAPR reduction performance. The proposed RC-SLM

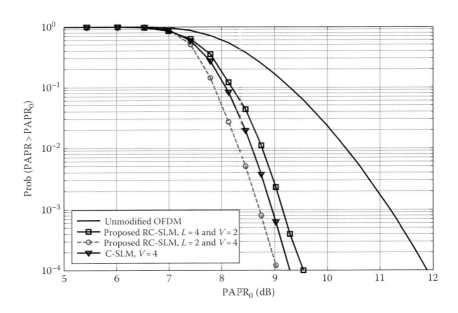

Figure 4.17 PAPR performance of the proposed RC-SLM scheme with various (L, V) combinations.

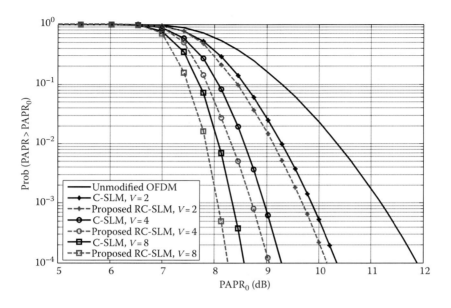

Figure 4.18 Comparison between the proposed RC-SLM scheme using $L = 2$ and the conventional SLM scheme in terms of CCDF.

scheme is shown to be able to achieve the lowest computational complexity when compared with [73,76] while at the same time improving the PAPR reduction performance by about 0.3 dB.

As in the conventional SLM scheme, the proposed RC-SLM scheme needs to transmit SI bits to indicate which phase sequence is selected at the transmitter to enable the receiver to recover the transmitted data blocks. In [52], we propose a small overhead (s-SLM) scheme which improves the detection probability of the SI bits and hence gives a better BER performance. We can apply this scheme to ensure correct detection of the SI bits. In this work, we assume that the SI bits can be correctly detected at the receiver.

4.9 Summary and Conclusions

This chapter proposed an s-SLM scheme for STBC-based MIMO-OFDM systems. The proposed s-SLM scheme was compared with other SLM schemes such as the SISO-SLM, i-SLM, and d-SLM schemes. It was demonstrated that the proposed scheme not only improves system bandwidth efficiency but also achieves a significantly superior BER performance than the other SLM schemes. The obtained results show that the detection probability of the SI bits in the proposed s-SLM scheme improves as a result of STD diversity, however incurring a slight degradation of the PAPR performance as compared to the i-SLM scheme. The PAPR performance of the proposed s-SLM scheme was improved by using an unequal power distribution strategy.

We also proposed a reduced complexity selective mapping (RC-SLM) scheme. The proposed scheme is based on partitioning the frequency domain symbol sequence into several sub-blocks to reduce the length of the phase sequences and then using a kind of low-complexity conversion to replace the IFFT stages. The performance of the proposed RC-SLM scheme along with unequal power distribution strategy was studied with computer simulation. The results obtained show that the proposed RC-SLM scheme is able to achieve the lowest computational complexity when compared with other low-complexity schemes proposed in the literature while at the same time improves the PAPR reduction performance by about 0.3 dB.

PART II

Performance Evaluation of the OFDM and SC-FDE Systems Using Continuous Phase Modulation

5.1 Introduction

As explained in Chapter 2, continuous phase modulation (CPM) can be used as a signal transformer to solve the peak-to-average power ratio (PAPR) problem. Using CPM, the high PAPR OFDM (orthogonal frequency division multiplexing) signal is transformed to a constant envelope signal (i.e., 0 dB PAPR). In this chapter, we first study the performance of the CPM-based OFDM (CPM-OFDM) system. Then we propose a CPM-based single-carrier frequency domain equalization (CPM-SC-FDE) structure for broadband wireless communication systems [18,80]. The proposed structure combines the advantages of the low complexity of SC-FDE, in addition to exploiting the channel frequency diversity and the power efficiency of CPM. Both the CPM-OFDM system and the proposed system are implemented with FDE to avoid the complexity of the equalization. Two types of frequency domain equalizers are considered and compared for performance evaluation of both systems; the zero forcing (ZF) equalizer and the minimum mean square error (MMSE) equalizer. Simulation experiments are performed for a variety of multipath fading channels.

The rest of this chapter is organized as follows. Section 5.2 presents the CPM-OFDM system model. The FDE process and the design of the equalizers are explained in Section 5.3. Section 5.4 presents the proposed CPM-SC-FDE system model. Section 5.5 presents the phase demodulator. In Section 5.6, we explain the spectral efficiency and the multipath diversity of CPM Signals. Section 5.7 provides the numerical results and discussion. Finally, Section 5.8 provides a summary and concludes this chapter.

5.2 CPM-OFDM System Model

A new approach to mitigate the PAPR problem is based on signal transformations. This approach involves a signal transformation prior to amplification, then an inverse transformation at the receiver prior to demodulation.

In [33,81–83], a phase modulator (PM) transform was considered to generate signals with CPM. This scheme is attractive for wireless communications because of the constant envelope of the generated signals, which is needed for power efficient transmitters, and its ability to exploit the diversity of the multipath channel, which is needed to improve the BER performance. In the CPM-OFDM system, the OFDM signal is used to phase-modulate the carrier. This system shares many of the same functional blocks with the conventional OFDM system. This makes the existing OFDM systems capable of providing an additional CPM-OFDM mode, easily.

The block diagram of the CPM-OFDM system is shown in Figure 5.1. Let $X(k)$ denote the M-ary quadrature amplitude modulation (QAM) data symbols. During each T-second block interval, an N_{DFT}-points IDFT (inverse discrete Fourier transform) is used to give the block of time samples $x(n)$ corresponding to $X(k)$. After this, the generated OFDM sequence, $x(n)$, passes through a phase modulator to obtain the constant envelope sequence $s(n) = \exp(jCx(n))$, where C is a scaling

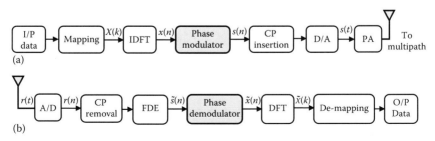

Figure 5.1   Block diagram of the CPM-OFDM system. (a) Transmitter and (b) receiver.

constant. After the PM, a cyclic prefix (CP) is added at the beginning of each data block to help mitigate the interblock interference (IBI), which is assumed to have a longer length than the channel impulse response.

The continuous-time CPM-OFDM signal $s(t)$ is then generated at the output of the digital-to-analog (D/A) converter. This CPM signal can be expressed as follows [18,33]:

$$s(t) = Ae^{j\phi(t)} = Ae^{j[2\pi h x(t)+\theta]}, \quad T_g \leq t < T \tag{5.1}$$

where

A is the signal amplitude

h is the modulation index

θ is an arbitrary phase offset used to achieve CPM [18,81]

T_g is the guard period

T is the block period

$x(t)$ is a real-valued OFDM message signal comprised of K subcarriers given as

$$x(t) = C_n \sum_{k=1}^{K} I_k q_k(t) \tag{5.2}$$

where I_k are the real-valued data symbols:

$$I_k = \begin{cases} \Re\{X(k)\}, & k \leq \dfrac{K}{2} \\ -\Im\left\{X\left(k-\dfrac{K}{2}\right)\right\}, & k > \dfrac{K}{2} \end{cases} \tag{5.3}$$

where

$\Re\{X(k)\}$, $\Im\{X(k)\}$ are the real and the imaginary part of $\{X(k)\}$, respectively

$q_k(t)$ is a function used to represent the orthogonal subcarriers and is expressed as follows:

$$q_k(t) = \begin{cases} \cos\left(\dfrac{2\pi k t}{T}\right), & k \leq \dfrac{K}{2} \\ \sin\left(\dfrac{2\pi(k-K/2)t}{T}\right), & k > \dfrac{K}{2} \end{cases} \tag{5.4}$$

Using equations (5.3) and (5.4), equation (5.2) can be rewritten as

$$x(t) = C_n \left[\sum_{k=1}^{K/2} \Re\{X(k)\} \cos\left(\frac{2\pi kt}{T}\right) - \sum_{k=K/2+1}^{K} \Im\{X(k)\} \sin\left(\frac{2\pi(k-K/2)t}{T}\right) \right]$$

(5.5)

where C_n is a normalization constant used to normalize the variance of the message signal (i.e., $\sigma_x^2 = 1$), and consequently the variance of the phase signal, $\sigma_\varphi^2 = (2\pi h)^2$. This requirement is achieved by setting C_n as follows [18]:

$$C_n = \sqrt{\frac{2}{K\sigma_I^2}}$$

(5.6)

where σ_I^2 is the variance of the data symbols. The assumption that the data is independent and identically distributed leads to

$$\sigma_I^2 = E\{|X(k)|^2\} = \frac{1}{M} \sum_{l=1}^{M} (2l-1-M)^2 = \frac{M^2-1}{3}$$

(5.7)

where M is the number of constellation points.

If J denotes the oversampling factor, there will be $N_{DFT} = JK$ samples per block, then with the help of equations (5.1) and (5.5), the discrete-time version of $s(t)$ at the output of the phase modulator can be expressed as

$$s(n) =$$

$$A \exp\left\{ j\left(2\pi h C_n \left[\sum_{k=1}^{K/2} I_k \cos\left(\frac{2\pi kn}{N_{DFT}}\right) + \sum_{k=K/2+1}^{K} I_k \sin\left(\frac{2\pi(k-K/2)n}{N_{DFT}}\right) \right] + \theta \right) \right\}$$

(5.8)

with $n = 0, 1, \ldots, JK - 1$.

The transmitted signal $s(t)$ then passes through the multipath channel. The channel impulse response is modeled as a wide-sense stationary uncorrelated scattering (WSSUS) process consisting of L discrete paths:

$$h(t) = \sum_{l=0}^{L-1} h(l)\delta(t - \tau_l)$$

(5.9)

where $h(l)$ and τ_l are the channel gain and delay of the lth path, respectively. The continuous-time received signal $r(t)$ is expressed as

$$r(t) = \sum_{l=0}^{L-1} h(l)s(t - \tau_l) + n(t) \tag{5.10}$$

where $n(t)$ is an additive white Gaussian noise (AWGN) with single-sided power spectral density N_0. The output of the A/D converter is sampled at a rate $f_s = JK/T$ sample/s. The nth ($n = -N_g, ..., 0, ..., N_{DFT} - 1$) sample of the received signal $r(t)$ is given by

$$r(n) = \sum_{i=0}^{LJ-1} h(i)s(n-i) + n(n) \tag{5.11}$$

where N_g is the number of samples in the guard interval and N_{DFT} is the number of samples per OFDM data block, as shown in Figure 5.2.

After the A/D, the CP samples are discarded and the remaining samples are equalized with a FDE process. Defining $N_{DFT} = JK$, the received signal $r(n)$ is transformed into the frequency domain using an N_{DFT}-points DFT. The received signal on the mth ($m = 0, 1, ..., N_{DFT} - 1$) subcarrier is given by

$$R(m) = H(m)S(m) + N(m) \tag{5.12}$$

where $R(m)$, $H(m)$, $S(m)$, and $N(m)$ are the N_{DFT}-points DFT of $r(n)$, $h(n)$, $s(n)$, and $n(n)$, respectively.

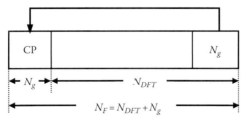

N_F: Total number of transmitted samples.

Figure 5.2 Transmitted data block.

5.3 Frequency Domain Equalizer Design

In this section, the design of the frequency domain equalizers is discussed. As shown in Figure 5.3 and explained in Chapter 2, the received signal is equalized in the frequency domain after the DFT block. The equalized signal is then transformed back into the time domain by using the IDFT.

Let $W(m)$, $(m = 0, 1, \ldots, N_{DFT} - 1)$, denote the equalizer coefficients for the mth subcarrier, the time domain equalized signal $\tilde{s}(n)$, which is the soft estimate of $s(n)$, can be expressed as follows:

$$\tilde{s}(n) = \frac{1}{N_{DFT}} \sum_{m=0}^{N_{DFT}-1} R(m)W(m)e^{j2\pi mn/N_{DFT}}, \quad n = 0, 1, \ldots, N_{DFT} - 1$$

(5.13)

The equalizer coefficients $W(m)$ are determined to minimize the mean squared error between the equalized signal $\tilde{s}(n)$ and the original signal $s(n)$. The equalizer coefficients are computed according to the type of the FDE as follows [11].

- The ZF equalizer:

$$W(m) = \frac{1}{H(m)}, \quad m = 0, 1, \ldots, N_{DFT} - 1 \qquad (5.14)$$

- The MMSE equalizer:

$$W(m) = \frac{H^*(m)}{\left|H(m)\right|^2 + (E_b/N_0)^{-1}} \qquad (5.15)$$

where $(.)^*$ denotes the complex conjugate. The ZF equalizer given in equation (5.14) perfectly eliminates the effect of the channel in the absence of noise, but when noise cannot be ignored, the ZF equalizer suffers from the noise enhancement phenomenon. On the other hand, the MMSE equalizer given in equation (5.15) takes into account the signal-to-noise ratio (SNR), making an optimum trade-off between the channel inversion and the noise enhancement.

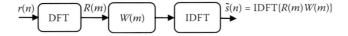

Figure 5.3   Frequency domain equalizer.

Considering the MMSE equalizer described in equation (5.15) and using equation (5.12), equation (5.13) can be rewritten as

$$\tilde{s}(n) = \underbrace{\frac{1}{N_{DFT}} \sum_{m=0}^{N_{DFT}-1} \frac{|H(m)|^2 S(m)}{|H(m)|^2 + (E_b/N_0)^{-1}} e^{j2\pi mn/N_{DFT}}}_{signal}$$

$$+ \underbrace{\frac{1}{N_{DFT}} \sum_{m=0}^{N_{DFT}-1} \frac{|H(m)|^* N(m)}{|H(m)|^2 + (E_b/N_0)^{-1}} e^{j2\pi mn/N_{DFT}}}_{noise} \quad (5.16)$$

Note that the MMSE and ZF equalizers are equivalent at high SNR. Considering the ZF equalizer, equation (5.16) can be rewritten as

$$\tilde{s}(n) = \underbrace{\frac{1}{N_{DFT}} \sum_{m=0}^{N_{DFT}-1} S(m) \, e^{j2\pi mn/N_{DFT}}}_{signal}$$

$$+ \underbrace{\frac{1}{N_{DFT}} \sum_{m=0}^{N_{DFT}-1} \frac{N(m)}{|H(m)|} e^{j2\pi mn/N_{DFT}}}_{noise} \quad (5.17)$$

5.4 Proposed CPM-SC-FDE System Model

SC-FDE [9–12,34–40] has found great popularity for application in wireless communication systems, especially for severe frequency-selective environments, due to its effectiveness and low complexity. Compared to OFDM, SC-FDE has a lower PAPR, less sensitivity to frequency synchronization errors, and a higher frequency diversity gain, when a relatively high rate channel coding scheme is applied.

In this section, the system model of the proposed CPM-SC-FDE structure is presented [18]. The block diagram of the CPM-SC-FDE system is shown in Figure 5.4. It is known from Chapter 2 that the main difference between the conventional OFDM system and SC systems is in the utilization of the DFT and IDFT operations. In OFDM systems, an IDFT block is placed at the

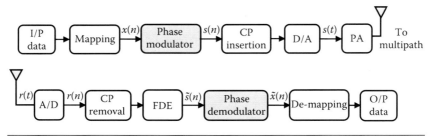

Figure 5.4   Block diagram of the proposed CPM-SC-FDE system.

transmitter to multiplex data into parallel subcarriers and a DFT block is placed at the receiver for FDE, while in SC systems both the DFT and IDFT blocks are placed at the receiver for FDE. When combined with DFT processing and the use of a CP, conventional OFDM systems and conventional SC-FDE systems are of equal complexity [9,10]. The case is different when using CPM where, as shown in Figures 5.1 and 5.4, the CPM-OFDM system needs two DFT operations and two IDFT operations. This makes the CPM-OFDM system more complex than the proposed CPM-SC-FDE system, which requires only a single DFT operation and a single IDFT operation for FDE.

Consider the transmission of a block of data in the proposed CPM-SC-FDE system over a multipath fading channel. Similar to the CPM-OFDM, the sequence $x(n)$ passes through a phase modulator to obtain a constant envelope sequence. Then CP is inserted between blocks to mitigate the IBI. In the CPM-OFDM case, the data symbols have an additional transformation by using the IDFT, $x(n) =$ IDFT$\{X(k)\}$, but in the SC-FDE case, no transformation is applied. At the receiver, the CP is removed and an FDE is performed similar to that in the CPM-OFDM system. Finally, phase demodulation and demapping are performed.

5.5 Phase Demodulator

In this section, the design of the phase demodulator is discussed. The phase demodulator block diagram is illustrated in Figure 5.5. It starts with a finite impulse response (FIR) filter to remove the out-of-band noise. The filter is designed using the windowing technique [84].

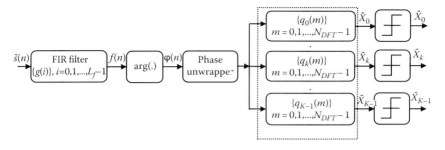

Figure 5.5 Phase demodulator.

If the filter impulse response has a length L_f and a normalized cutoff frequency f_{nor} ($0 < f_{nor} \leq 1$), it can be expressed as follows [18]:

$$g(i) = \frac{\sin\left(2\pi f_{nor}\left(i - \frac{L_f - 1}{2}\right)\right)}{\pi\left(i - \frac{L_f - 1}{2}\right)}, \quad 0 \leq i \leq L_f - 1 \qquad (5.18)$$

In equation (5.18), if $i = (L_f - 1)/2$, $g(i) = 2\pi f_{nor}/\pi$. The output of the FIR filter can be expressed as

$$f(n) = \sum_{i=0}^{L_f - 1} g(i)\tilde{s}(n - i) \qquad (5.19)$$

Afterward, the phase of the filtered signal $f(n)$ is obtained:

$$\varphi(n) = \arg\left(f(n)\right) = \phi(n) + \delta(n) \qquad (5.20)$$

where
  $\phi(n)$ denotes the phase of the desired signal
  $\delta(n)$ denotes the phase noise

Then, a phase unwrapper is used to minimize the effect of any phase ambiguities and to make the receiver insensitive to phase offsets caused by channel nonlinearities.

Finally, a bank of K matched filters is used to obtain the soft estimates of the data symbols, which are then passed through a group of decision devices to obtain hard estimates of the data.

5.6 Spectral Efficiency and Multipath Diversity of CPM Signals

The spectral efficiency is an important quality metric for a modulation format, since it quantifies how many information bits per second can be loaded per unity of the available bandwidth. To evaluate the spectral efficiency of a signal, its bandwidth needs to be estimated. Using Taylor expansion, the CPM signal described in equation (5.1), when $\theta = 0$, can be rewritten as

$$s(t) = Ae^{j2\pi hx(t)} = A\sum_{n=0}^{\infty}\left[\frac{(j2\pi h)^n}{n!}\right]x^n(t)$$

$$= A\left[1 + j2\pi hx(t) - \frac{(2\pi h)^2}{2!}x^2(t) - j\frac{(2\pi h)^3}{3!}x^3(t) + \cdots\right] \quad (5.21)$$

The subcarriers are centered at the frequencies $\pm i/T$ Hz, $i = 1$, $2, \ldots, K/2$. The effective double-side bandwidth of the message signal, $x(t)$, is defined as $W = K/T$ Hz. According to equation (5.21), the bandwidth of $s(t)$ is at least W, if the first two terms only of the summation are considered. Depending on the modulation index value, the effective bandwidth can be greater than W. A useful bandwidth expression for the CPM signal is the root-mean-square (RMS) bandwidth [85]:

$$\text{BW} = \max(2\pi h, 1)W \quad \text{Hz} \quad (5.22)$$

As shown in equation (5.22), the signal bandwidth grows with $2\pi h$, which in turn reduces the spectral efficiency. Since the bit rate is $R = K(\log_2 M)/T$ bps, the spectral efficiency of the CPM signal, η, can be expressed as

$$\eta = \frac{R}{\text{BW}} = \frac{\log_2 M}{\max(2\pi h, 1)} \quad \text{bps/Hz} \quad (5.23)$$

The spectral efficiency of a CPM signal is controlled by two parameters, M and $2\pi h$. On the other hand, the spectral efficiency of an OFDM signal is $\log_2 M$, which depends only on M.

The Taylor expansion given in equation (5.21) reveals how a CPM signal exploits the frequency diversity in the channel for a large modulation index. This is not necessarily the case, however. For a small modulation index, only the first two terms in equation (5.21) contribute:

$$s(t) \approx A\left[1 + j2\pi h m(t)\right] \qquad (5.24)$$

In this case, the CPM signal does not have the frequency spreading given by the higher-order terms. Therefore, the CPM signal does not have the ability to exploit the frequency diversity of the channel.

5.7 Numerical Results and Discussion

In this section, simulation experiments are carried out to demonstrate the performance of the CPM-OFDM system and the proposed CPM-SC-FDE system. Two types of frequency domain equalizers are used in these experiments: the ZF equalizer and the MMSE equalizer. A 4-ary ($M = 4$) QAM is used in the simulations. Each block contains $K = 64$ symbols and each symbol is sampled eight times ($J = 8$). The channel is assumed to be perfectly known at the receiver. The SNR is defined as the ratio between the average received signal power and the noise power. The FIR filter has an impulse response length of $L_f = 11$ and a normalized cutoff frequency of $f_{nor} = 0.2$ [18,33]. Simulations are performed using four frequency-selective fading channel models as follows:

1. *Channel A*, which has a weak secondary path (one-tenth, i.e., –10 dB the power of the primary path).
2. *Channel B*, which has a strong secondary path (one-half, i.e., –3 dB the power of the primary path).
3. *Channel C*, which has an exponential delay power spectral density:

$$E\left\{|b_l|^2\right\} = \frac{1}{C_e} \exp\left(\frac{-\tau_l}{\tau_{rms}}\right), \quad 0 \le \tau_l \le \tau_{max} \ \mu s \qquad (5.25)$$

where
$$C_e = \sum_{l=0}^{N_C} \exp(-\tau_l/\tau_{rms})$$

N_c represents the number of channel taps
τ_{max} is the channel maximum propagation delay

4. *Channel D*, which has a uniform power spectral density:

$$E\left\{\left|h_l\right|^2\right\} = \frac{1}{N_c}, \quad 0 \le \tau_l \le \tau_{max} \quad \mu s \qquad (5.26)$$

For each of the considered channel models, the guard period is of sufficient duration, $T_g > \tau_{max}$.

Figure 5.6 shows the effect of the modulation index on the performance of the CPM-OFDM system, at a fixed SNR = 20 dB, for both single- and multipath channel cases. It is clear from this figure that the BER in the CPM-OFDM system decreases with the increase of the modulation index $2\pi h$ and reaches its optimum value at $2\pi h = 1.3$. It is also clear that the performance of the CPM-OFDM system over a multipath channel is better than its performance over a single-path channel for $2\pi h > 0.4$, which clarifies that the proposed CPM-OFDM system exploits the multipath diversity of the channel for large modulation indices. For small modulation indices, between 0 and 0.4, the performance over a single-path channel constitutes a lower bound for the performance over a multipath channel as indicated by Figure 5.7.

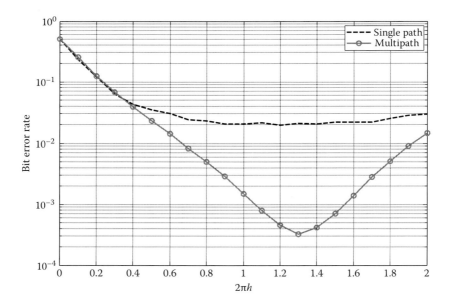

Figure 5.6   Impact of the modulation index on the performance of the CPM-OFDM system for both single-path and multipath channel cases at SNR = 20 dB.

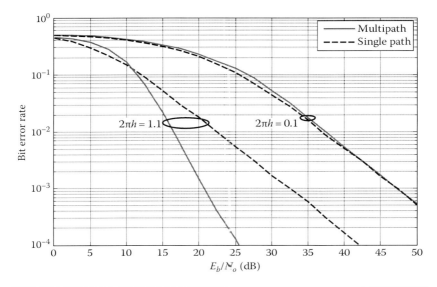

Figure 5.7 BER performance of the CPM-OFDM system over single-path and multipath channels, using the channel model *C* and an MMSE equalizer.

From this discussion it is clear that there is a trade-off between the BER performance which is improved by increasing the modulation index and the spectral efficiency which is improved by decreasing the modulation index. Based on the performance shown in Figures 5.6 and 5.7, and the spectral efficiency given by equation (5.23), it can be deduced that a moderate modulation index leads to an efficient utilization of the channel frequency diversity, while maintaining an acceptable spectral efficiency. In the following results, we will use $2\pi h = 1$ to make a good trade-off between the BER performance and spectral efficiency.

Figures 5.8 and 5.9 show a comparison in performance between the CPM-OFDM system and the proposed CPM-SC-FDE system over the four channel models discussed that use the ZF and MMSE equalizers. The performance over a Rayleigh frequency-nonselective fading channel ($L = 1$) is also plotted. The results show that the MMSE equalizer is better than the ZF equalizer in all cases, due to the noise enhancement effect resulting from the ZF equalizer. The results also show the impact of the multipath diversity. In the CPM-OFDM system, the performance over channels A to D is better than the performance over the frequency-nonselective Rayleigh ($L = 1$) channel for SNR > 15 dB if an MMSE equalizer is used. On the other hand, the

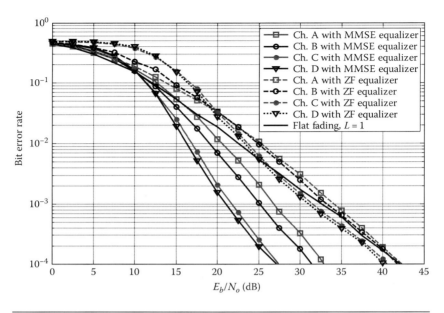

Figure 5.8 Performance of the CPM-OFDM system using a ZF equalizer and an MMSE equalizer.

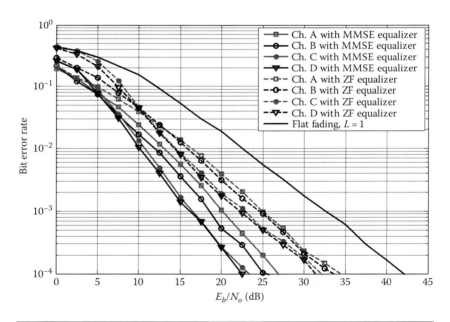

Figure 5.9 Performance of the proposed CPM-SC-FDE system using a ZF equalizer and an MMSE equalizer.

performance of the proposed CPM-SC-FDE system over channels A to D is better than the performance over the flat fading channel for both types of equalizers at all values of the SNR. The results also show that the performance of both systems improves as the number of channel multipaths increases.

Figure 5.10 shows a comparison in the performance between the CPM-OFDM system, the proposed CPM-SC-FDE system, and the conventional OFDM system. It is clear from this figure that at a BER of 10^{-3}, the proposed CPM-SC-FDE system outperforms the CPM-OFDM system by 5 dB, which is a significant improvement when compared to the SC-FDE system presented in [86–88]. Both systems have better performance than that of the conventional OFDM system. The figure reveals that the proposed CPM-SC-FDE system solves the problems of the CPM-OFDM system at low SNRs.

Figure 5.11 shows a comparison between the proposed CPM-SC-FDE system and the CPM-OFDM system in the effect of the modulation index on the system performance for transmission over a multipath channel. It is clear that the proposed CPM-SC-FDE system outperforms the CPM-OFDM system at relatively small

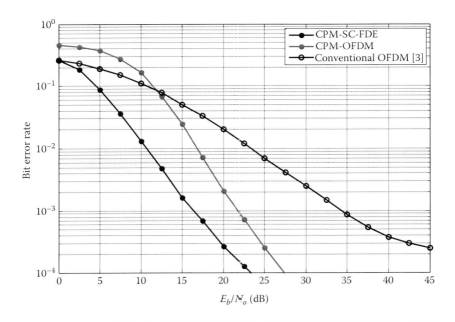

Figure 5.10   BER performance of the conventional OFDM system, the CPM-OFDM system, and the proposed CPM-SC-FDE system using channel model C and an MMSE equalizer.

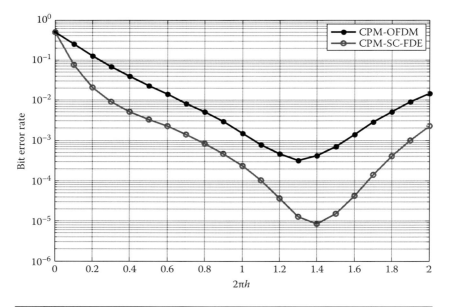

Figure 5.11 Effect of the modulation index on the performance of the proposed CPM-SC-FDE system and the CPM-OFDM system for transmission over channel model C with a MMSE equalizer at SNR = 20 dB.

modulation index values. To obtain the same BER performance, the proposed CPM-SC-FDE requires a smaller modulation index than the CPM-OFDM system. Therefore, we can say that the proposed CPM-SC-FDE system has better spectral efficiency than the CPM-OFDM system. For example, at $M = 4$ and BER of 3.3×10^{-4}, the CPM-SC-FDE system needs $2\pi h = 0.96$, and according to equation (5.23), the spectral efficiency will be $\eta = 2$ bps. On the other hand, the CPM-OFDM system needs $2\pi h = 1.3$, that is, $\eta = 1.54$ bps, which means that the CPM-SC-FDE system can achieve 30% improvements in spectral efficiency when compared to the CPM-OFDM system.

5.8 Summary and Conclusions

In this chapter, a novel CPM-SC-FDE system is presented and its performance compared to the CPM-OFDM system and the conventional OFDM system. It is observed that both CPM-based systems give a better BER performance than the conventional OFDM system. The effect of the modulation index on both CPM-based systems has been studied. The simulation experiments have shown that both CPM-based systems exploit the frequency diversity of multipath channels. The proposed CPM-SC-FDE system gives a better performance, better utilization of the channel frequency diversity, and better spectral efficiency than the CPM-OFDM system. A trade-off can be made between the BER performance and the spectral efficiency in the proposed CPM-SC-FDE system by appropriate selection of the modulation index value.

The obtained numerical results showed that the performance over the multipath channels is at least 5 and 12 dB better than the performance over the single-path channel using the CPM-OFDM and CPM-SC-FDE systems, respectively. The results also show that when CPM is used, SC-FDE systems can outperform OFDM systems by about 5 dB. Finally, the CPM-SC-FDE system can achieve 30% improvements in spectral efficiency when compared to the CPM-OFDM system.

PART III

6

CHAOTIC INTERLEAVING SCHEME FOR THE CPM-OFDM AND CPM-SC-FDE SYSTEMS

6.1 Introduction

As explained in Chapter 5, continuous phase modulation (CPM) is widely used in wireless communication systems, because of the constant envelope of the transmitted signals, which is required for efficient power transmission, and its ability to exploit the diversity of the multipath channel, which is needed to improve the bit error rate (BER) performance. However, CPM signals usually do not provide a high spectral efficiency when compared with non-constant-envelope modulations such as pulse amplitude modulation (PAM) and quadrature amplitude modulation (QAM) [89].

Strong mechanisms for error reduction such as powerful error correction codes [90] and efficient interleaving schemes [91] are required to reduce the channel effects on the data transmitted. Since the channel errors caused by the mobile wireless channels are bursty in nature, interleaving is a must in mobile communication systems. Several interleaver schemes have been proposed. The simplest and most popular of such schemes is the block interleaver scheme [91,92]. In spite of the success of this scheme in achieving a good performance in wireless communication systems, there is a need for a much powerful scheme for severe channel degradation cases. Chaotic maps have been proposed for a wide range of applications in communications [93] and cryptography [94–97]. Due to the inherent strong randomization ability of these maps, they can be efficiently used for data interleaving.

In this chapter, we propose a chaotic interleaving scheme for the CPM-OFDM (orthogonal frequency division multiplexing) system and the CPM-SC-FDE (single-carrier frequency domain equalization) system [18,98–101]. Chaotic interleaving is used to generate permuted versions from the sample sequences to be transmitted, with low correlation among their samples, and hence a better BER performance can be achieved. The proposed CPM-OFDM and CPM-SC-FDE systems with chaotic interleaving combine the advantages of the frequency diversity and the high power efficiency of CPM-based systems with performance improvements due to chaotic interleaving. The BER performance of both systems with and without chaotic interleaving is evaluated by computer simulations. In addition, a comparison is made between chaotic interleaving and block interleaving. Simulation results show that the proposed chaotic interleaving scheme can greatly improve the performance of the CPM-OFDM and CPM-SC-FDE systems. Furthermore, the results show that the chaotic interleaving scheme outperforms the traditional block interleaving scheme in both systems. The results also show that the use of chaotic interleaving with the CPM-OFDM and CPM-SC-FDE systems provides a good trade-off between system performance and spectral efficiency.

The rest of this chapter is organized as follows. Section 6.2 presents the proposed CPM-OFDM system model with chaotic interleaving. The proposed CPM-SC-FDE system with chaotic interleaving is presented in Section 6.3. The block interleaving and the proposed chaotic interleaving schemes are explained in Section 6.4. Section 6.5 presents the equalizer design. Section 6.6 introduces the simulation results. Finally, Section 6.7 gives concluding remarks.

6.2 Proposed CPM-OFDM System with Chaotic Interleaving

A block diagram of the proposed CPM-OFDM system is shown in Figure 6.1 [100]. A block length of K symbols is assumed with $X(k)$ ($k = 0, 1, \ldots, K - 1$) representing the data sequence after symbol mapping and $x(n)$ ($n = 0, 1, \ldots, N_{DFT} - 1$) representing the N_{DFT} point IDFT (inverse discrete Fourier transform) of the data sequence $X(k)$. During each T-second symbol interval, the time samples $x(n)$ are subjected to a phase modulation (PM) step to get the constant envelope

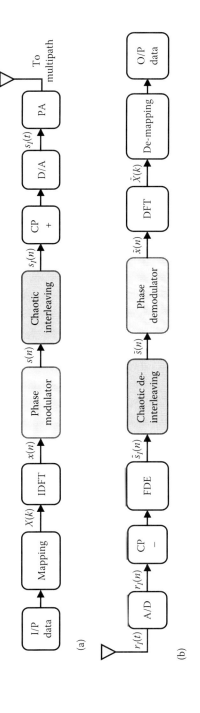

Figure 6.1   CPM-OFDM with chaotic interleaving system model. (a) Transmitter and (b) receiver.

sequence, $s(n)$. After the PM, interleaving is applied to the phase-modulated samples. The obtained sequence is $s_I(n)$, with the subscript I referring to the interleaving process. Then each data block is padded with a CP to mitigate interblock interference (IBI). The CP length must be longer than the channel impulse response. Finally, the continuous-time signal, $s_I(t)$, is generated at the output of the digital-to-analog converter.

6.3 Proposed CPM-SC-FDE System with Chaotic Interleaving

The block diagram of the proposed CPM-SC-FDE system is shown in Figure 6.2 [101]. It is known from Chapter 2 that the main difference between the conventional OFDM system and SC systems is in the utilization of the DFT and IDFT operations. When combined with DFT processing and the use of a CP, conventional OFDM systems and conventional SC-FDE systems are of equal complexity [9,10]. The situation is different when using CPM, where, as shown in Figures 6.1 and 6.2, the CPM-OFDM system needs two DFT operations and two IDFT operations. This makes the CPM-OFDM system more complex than the CPM-SC-FDE system, which requires only a single DFT operation and a single IDFT operation for FDE.

As explained in Chapter 5, $s_I(t)$ can be written as

$$s_I(t) = Ae^{j\phi(t)} = Ae^{j[2\pi h x(t)+\theta]}, \quad T_g \leq t < T \tag{6.1}$$

As will be shown in the next sections, the proposed modifications in this chapter will be in the interleaver and equalizer blocks.

6.4 Interleaving Mechanisms

Error correction codes are usually used to protect signals through transmission over wireless channels. Most of the error correction codes are designed to correct random channel errors. However, channel errors caused by mobile wireless channels are bursty in nature. Interleaving is a process to rearrange the samples of the transmitted signal so as to spread bursts of errors over multiple code words. The simplest and most popular of such interleavers is the block interleaver. We first review the basics of block interleaving [91] and then present the proposed chaotic interleaving mechanism.

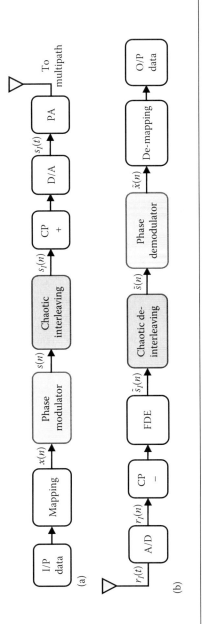

Figure 6.2 CPM–SC–FDE with chaotic interleaving system model. (a) Transmitter and (b) receiver.

6.4.1 Block Interleaving Mechanism

The idea of block interleaving can be explained with the aid of Figure 6.3. After PM, block interleaving is applied to the signal samples. The samples are first arranged in a matrix in a row-by-row manner and then read from this matrix in a column-by-column manner. Now let us look at how the block interleaving mechanism can correct error bursts. Assume a burst of errors affecting four consecutive samples (one-dimensional [1-D] error burst) as shown in Figure 6.3b with shades. After de-interleaving, the error burst is effectively spread among four different rows, resulting in a small effect for the 1-D error burst, as shown in Figure 6.3c. With a single error correction capability, it is obvious that no decoding error will result from the presence of such 1-D error burst. This simple example demonstrates the effectiveness of the block interleaving mechanism in combating 1-D bursts of errors. Let us examine the performance of the block interleaving mechanism when a two-dimensional (2-D) (2 × 2) error burst occurs [91], as shown in Figure 6.3b with shades. Figure 6.3c indicates that this 2 × 2 error burst has not been spread, effectively, and that there are adjacent samples in error in the first and the second rows. As a result, this error burst cannot be corrected using a single error correction mechanism. That is, the block interleaving mechanism cannot combat the 2 × 2 burst of errors.

6.4.2 Proposed Chaotic Interleaving Mechanism

As mentioned above, block interleavers are not efficient with 2-D bursts of errors. As a result, there is a need for advanced interleavers for this task. The 2-D chaotic Baker map in its discretized version is a good candidate for this purpose. After PM, the signal samples can be arranged to a 2-D format and then randomized using the chaotic Baker map. The chaotic interleaver generates permuted sequences with lower correlation between their samples and adds a degree of encryption to the transmitted signal.

The discretized Baker map is an efficient tool to randomize the items in a square matrix. Let $B(n_1, \ldots, n_k)$ denote the discretized map, where the vector $[n_1, \ldots, n_k]$, represents the secret key, S_{key}. Defining N as the number of data items in one row, the secret key is chosen such that each integer n_i divides N, and $n_1 + \cdots + n_k = N$.

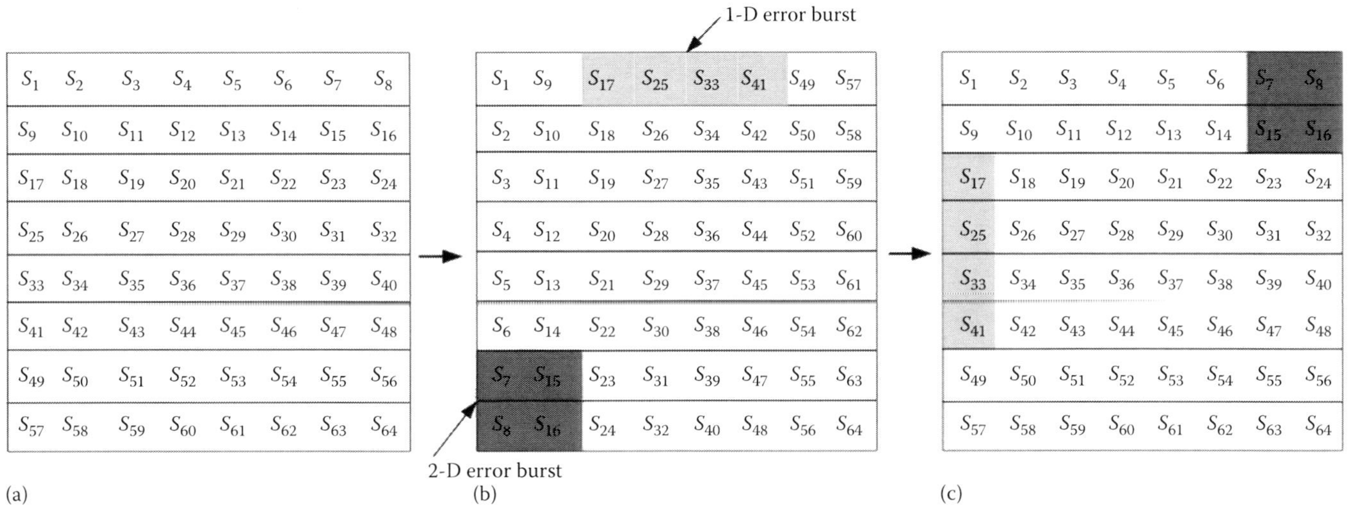

Figure 6.3 Block interleaving of an 8 × 8 matrix. (a) The 8 × 8 matrix, (b) block interleaving of the 8 × 8 matrix, and (c) effect of error bursts after the de-interleaving.

Let $N_i = n_1 + \ldots + n_i$. The data item at the indices (q, z), is moved to the indices [96,101]

$$
B_{(n_1,\ldots,n_k)}(q,z) = \left(\frac{N}{n_i}(q - N_i) + z \bmod \left(\frac{N}{n_i} \right), \frac{n_i}{N} \left(z - z \bmod \left(\frac{N}{n_i} \right) \right) + N_i \right)
$$

(6.2)

where $N_i \leq q < N_i + n_i$, and $0 \leq z < N$.

Chaotic permutation is performed using the following steps [99]:

1. An $N \times N$ square matrix is divided into k rectangles of width n_i and number of elements N.
2. The elements in each rectangle are rearranged to a row in the permuted rectangle. Rectangles are taken from right to left beginning with the upper rectangles then the lower ones.
3. Inside each rectangle, the scan begins from the bottom left corner toward the upper elements.

Figure 6.4 shows an example for the chaotic interleaving of an (8×8) square matrix (i.e., $N = 8$). The secret key $S_{key} = [n_1, n_2, n_3] = [2, 4, 2]$. Note that the chaotic interleaving mechanism has a better treatment to both 1-D and 2-D bursts of errors than the block interleaving mechanism. Errors are better distributed to samples after de-interleaving in the proposed chaotic interleaving mechanism. As a result, a better BER performance can be achieved with the proposed interleaving mechanism for the same error correcting code. In addition, it adds a degree of security to the communication system. At the receiver of the proposed systems with chaotic interleaving, the received signal is passed through an analog-to-digital converter, then the CP is discarded and the remaining samples are equalized.

6.5 Equalizer Design

The design of frequency domain equalizers was explained in Chapter 5. In this section we define a new type of equalizer: the regularized zero forcing (RZF) equalizer. As shown in Figures 6.1 and 6.2, the received signal is equalized in the frequency domain. Let $W(m)$, $(m = 0, 1, \ldots, N_{DFT} - 1)$, denote the equalizer coefficients

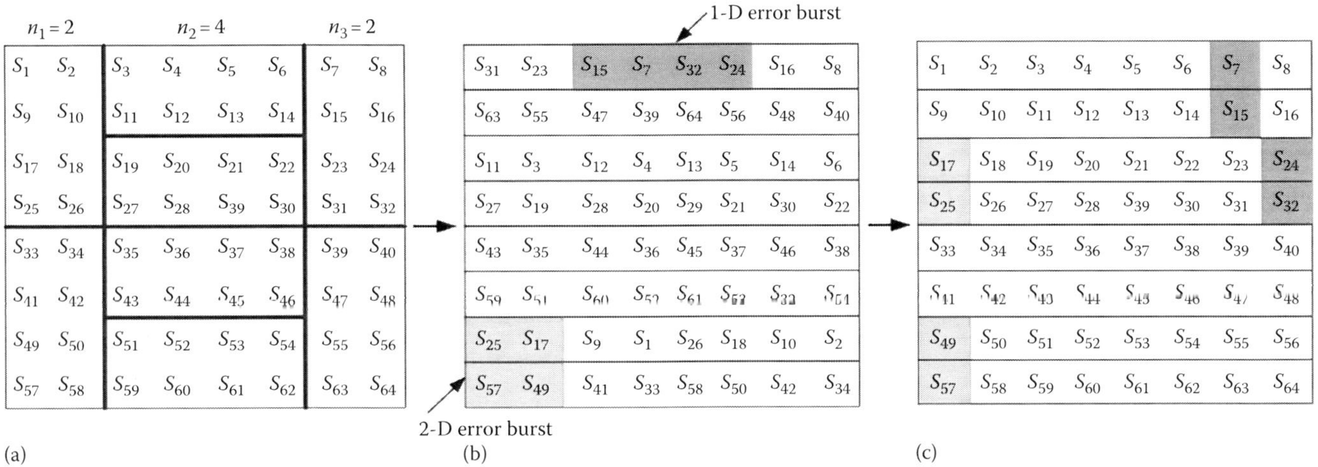

Figure 6.4 Chaotic interleaving of an 8×8 matrix. (a) The 8×8 matrix divided into square rectangles, (b) chaotic interleaving of the 8×8 matrix, and (c) effect of error bursts after the de-interleaving.

for the mth subcarrier, then the time domain equalized signal $\tilde{s}_I(n)$, which is the soft estimate of $s_I(n)$, can be expressed as follows:

$$\tilde{s}_I(n) = \frac{1}{N_{DFT}} \sum_{m=0}^{N_{DFT}-1} W(m)R_I(m)e^{j2\pi mn/N_{DFT}} \tag{6.3}$$

The equalizer coefficients $W(m)$ are selected to minimize the mean squared error between the equalized signal $\tilde{s}_I(n)$ and the original signal $s_I(n)$. The equalizer coefficients are computed according to the type of the FDE. The zero forcing (ZF) and minimum mean square error (MMSE) equalizer formulas are given in Chapter 5. The RZF equalizer is defined by

$$W(m) = \frac{H^*(m)}{|H(m)|^2 + \beta} \tag{6.4}$$

where
  $(.)^*$ denotes the complex conjugate
  β is the regularization parameter

The RZF equalizer described in equation (6.4) avoids the problems associated with the MMSE equalizer, such as the measurement of the signal power and the noise power, which are not available prior to equalization. In addition, the RZF equalizer avoids the noise enhancement caused by the ZF equalizer by introducing the regularization parameter β into the equalization process.

De-interleaving is then applied to the equalized samples. After this, a phase demodulation step is applied to recover the data, as explained in Chapter 5.

6.6 Numerical Results and Discussion

In this section, simulation experiments are carried out to evaluate the performance of the CPM-OFDM and CPM-SC-FDE systems, when chaotic interleaving is applied in both systems, using the simulation parameters described in Chapter 5. A channel model following the exponential delay profile in [18] (channel model C) with a root mean square delay spread $\tau_{rms} = 2$ μs is adopted, except in Figure 6.9. The channel is assumed to be perfectly known at the receiver.

6.6.1 Proposed CPM–OFDM System Results

Figure 6.5 demonstrates the relation between the regularization parameter β and the BER for the proposed CPM-OFDM system with chaotic interleaving at different SNR values. According to this figure, the best choice of β is 10^{-3}.

Figure 6.6 shows the BER performance of the proposed CPM-OFDM system with chaotic interleaving using ZF, RZF (with β = 10^{-3}), and MMSE frequency domain equalizers. The results show that the MMSE equalizer outperforms both ZF and RZF. For example, at BER = 10^{-2}, MMSE outperforms RZF and ZF by 0.7 and 8 dB, respectively. On the other hand, the RZF avoids the problems associated with the MMSE, such as the measurement of signal power and noise power, which are not available prior to equalization, with a slight degradation in SNR. Moreover, the RZF avoids the noise enhancement caused by the ZF equalizer by the proper selection of the regularization parameter β.

Figure 6.7 shows the performance comparison between the proposed CPM-OFDM system and the CPM-OFDM system described in [18] in terms of modulation index using an MMSE equalizer. It is clear that the proposed CPM-OFDM system outperforms the conventional CPM-OFDM system at a relatively small modulation

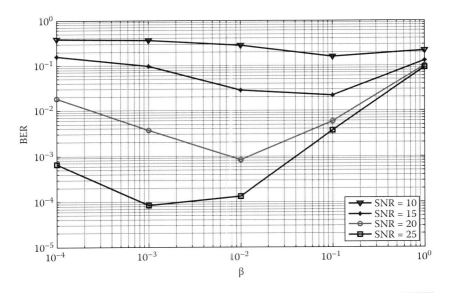

Figure 6.5   BER versus the regularization parameter at different SNRs.

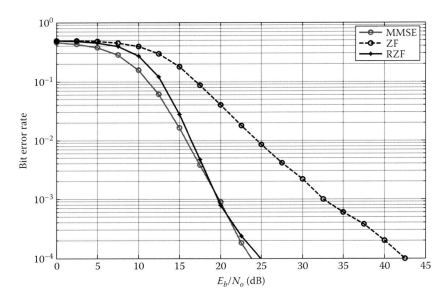

Figure 6.6 BER performance of the CPM-OFDM system using ZF, RZF, and MMSE-FDEs.

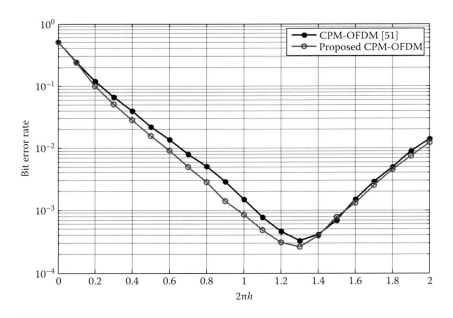

Figure 6.7 BER performance of the conventional CPM-OFDM and the proposed CPM-OFDM systems in terms of $2\pi h$ at SNR = 20 dB.

index. So we can say that the chaotic interleaving scheme improves the spectral efficiency when used with the CPM-OFDM system.

Table 6.1 shows the improvement percentage (**G**) in spectral efficiency of the proposed CPM-OFDM system, with chaotic interleaving over the conventional CPM-OFDM system [18]. For example, at BER of 3.2×10^{-4} and $M = 4$, the proposed CPM-OFDM system needs $2\pi h = 1.17$ and according to equation (5.23), the spectral efficiency will be $\eta = 1.71$ bps. On the other hand, the conventional CPM-OFDM system needs $2\pi h = 1.3$, that is, $\eta = 1.53$ bps, which means that the proposed CPM-OFDM system with chaotic interleaving can achieve about 12% improvements in spectral efficiency when compared to the conventional CPM-OFDM system.

Figure 6.8 presents a comparison between the performance of the conventional CPM-OFDM system without interleaving [18], the

Table 6.1 Improvement Percentage (**G**) in Spectral Efficiency in the CPM-OFDM System with and without Chaotic Interleaving

BER	CPM-OFDM (NO INTERLEAVING)	CPM-OFDM (WITH CHAOTIC INTERLEAVING)	**G** (%)
8×10^{-4}	$\eta = 1.818$ bps ($2\pi h = 1.1$)	$\eta = 2$ bps ($2\pi h = 1$)	10
3.2×10^{-4}	$\eta = 1.53$ bps ($2\pi h = 1.3$)	$\eta = 1.71$ bps ($2\pi h = 1.17$)	12

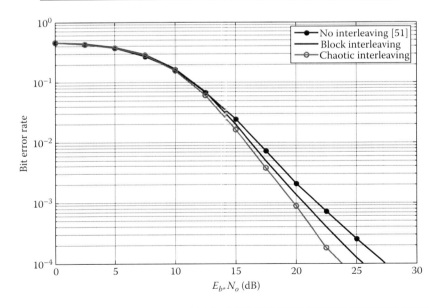

Figure 6.8 BER performance of CPM-OFDM with no interleaving, with block interleaving, and with the proposed chaotic interleaving.

CPM-OFDM system with block interleaving, and the CPM-OFDM system with the proposed chaotic interleaving. It is clear that the proposed CPM-OFDM system outperforms both the conventional CPM-OFDM and the CPM-OFDM with block interleaving. For example, at BER = 10^{-3}, the proposed CPM-OFDM system provides SNR improvements of 2.3 and 1.1 dB over the conventional CPM-OFDM system [18] and the CPM-OFDM system with block interleaving, respectively. This improvement is attributed to the strong randomization ability of the proposed chaotic interleaver.

Figure 6.9 shows the performance of the conventional OFDM, the conventional CPM-OFDM [18], and the proposed CPM-OFDM systems in terms of a BER versus the normalized RMS delay spread, at a fixed SNR = 20 dB. In the flat fading case (i.e., $\tau_{rms} = 0$), the performance of the conventional CPM-OFDM and the proposed CPM-OFDM systems converge with a small performance loss compared to the conventional OFDM system, due to the effect of the phase demodulator threshold. In frequency-selective channels ($\tau_{rms} > 0$), however, the proposed CPM-OFDM system achieves a significant performance improvement over the other systems. It also provides a better performance than that in the flat fading case by achieving frequency diversity, especially at high delay spreads.

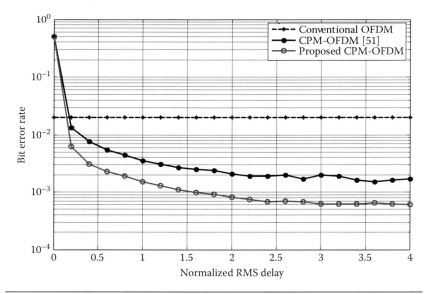

Figure 6.9 Effect of the RMS delay on the performance of the proposed CPM-OFDM, the conventional CPM-OFDM, and the conventional OFDM systems at an SNR = 20 dB.

6.6.2 *Proposed CPM–SC–FDE System Results*

Figure 6.10 evaluates the effect of the choice of the regularization parameter on the proposed CPM–SCFDE system with chaotic interleaving. The variation of the BER with the regularization parameter is clear in this figure at different SNR values. The objective of this figure is to choose an optimum value for the regularization parameter if an RZF equalizer is to be used in the proposed system. According to this figure, the best choice of β is 10^{-2}.

Figure 6.11 shows the BER performance of the proposed CPM–SC–FDE system with chaotic interleaving using the ZF equalizer, the RZF equalizer (with $\beta = 10^{-2}$), and the MMSE equalizer. The results show that the MMSE equalizer outperforms both the ZF equalizer and the RZF equalizer. For example, at a BER = 10^{-3}, the MMSE equalizer outperforms the RZF equalizer and the ZF equalizer by about 0.3 and 9.5 dB, respectively.

Figure 6.12 shows a performance comparison between the conventional SC–FDE system [9], the CPM–SC–FDE system without interleaving [18], the CPM–SC–FDE system with block interleaving, and the proposed CPM–SC–FDE system with chaotic interleaving using the MMSE equalizer in all systems. It is clear that the proposed

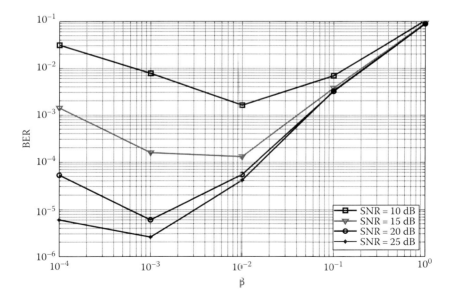

Figure 6.10 Proposed CPM-SC-FDE BER versus the regularization parameter at different SNRs.

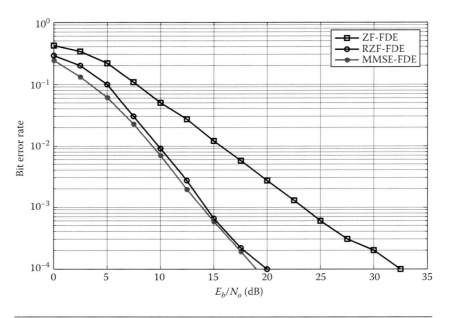

Figure 6.11   BER performance of the proposed CPM-SC-FDE system with chaotic interleaving using the ZF equalizer, the RZF equalizer, and the MMSE equalizer.

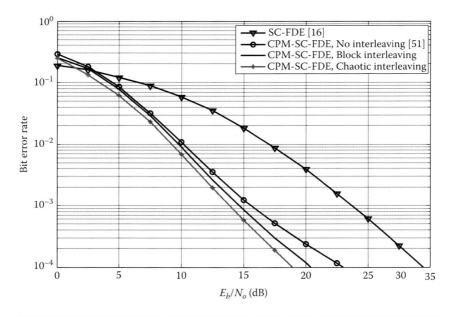

Figure 6.12   BER performance of the SC-FDE system, the CPM-SC-FDE system without interleaving, the CPM-SC-FDE system with block interleaving, and the proposed CPM-SC-FDE system with chaotic interleaving.

CPM-SC-FDE system with chaotic interleaving outperforms all the other systems. For example, at a BER = 10⁻³, the proposed CPM-SC-FDE system with chaotic interleaving provides SNR gains of 2 and 1 dB over the CPM-SC-FDE system without chaotic interleaving and the CPM-SC-FDE system with block interleaving, respectively.

Figures 6.13 and 6.14 show the effect of the modulation index on the performance of the CPM-SC-FDE system, at a fixed SNR = 20 dB for both the single- and the multipath cases with and without chaotic interleaving, respectively. In both cases the performance of the CPM-SC-FDE system in multipath channels outperforms in the single-path channel for large modulation index values, which confirms the analysis in Chapter 5. In multipath channels, the performance of the CPM-SC-FDE system with and without chaotic interleaving is better than its performance in the single-path channel for $2\pi h > 0.2$ and $2\pi h > 0.4$, respectively. Based on the performance results shown in Figure 6.14 and the spectral efficiency given by equation (5.23), it can be deduced that a moderate value of the modulation index achieves a significant utilization of the frequency diversity while maintaining a high spectral efficiency.

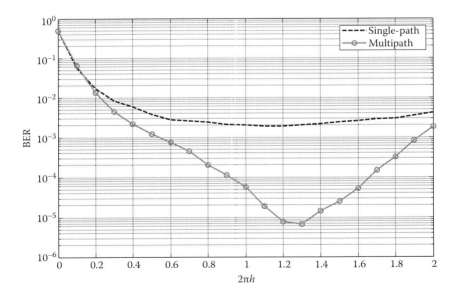

Figure 6.13 Impact of the modulation index on the performance of the CPM-SC-FDE system with chaotic interleaving for both the single- and multipath cases using an MMSE equalizer at an SNR = 20 dB.

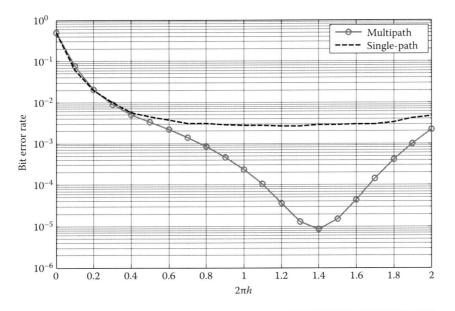

Figure 6.14 Impact of the modulation index on the performance of the CPM-SC-FDE system without interleaving for both the single- and multipath cases using an MMSE equalizer at an SNR = 20 dB.

Figure 6.15 shows the comparison between the performance of the CPM-SC-FDE with and without chaotic interleaving in terms of the modulation index in the multipath channel case. It is clear that the proposed CPM-SC-FDE system with chaotic interleaving outperforms the CPM-SC-FDE system without interleaving at relatively small modulation index values. This means that the proposed chaotic interleaving scheme improves the spectral efficiency in the CPM-SC-FDE system.

Table 6.2 shows the percentage of improvement in the spectral efficiency in the proposed CPM-SCFDE system with chaotic interleaving over the CPM-SC-FDE system without interleaving when $M = 4$. For example, at a BER of 8×10^{-6}, the CPM-SC-FDE system without interleaving needs $2\pi h = 1.4$, which gives a spectral efficiency of $\eta = 1.43$ bps. On the other hand, the proposed CPM-SC-FDE system with chaotic interleaving needs $2\pi h = 1.2$, that is, $\eta = 1.67$ bps, which means that the chaotic interleaving achieves an improvement in spectral efficiency of about 16%, when it is used with the CPM-SC-FDE system. We can conclude that the proposed CPM-SC-FDE system with chaotic interleaving achieves a trade-off between performance and spectral efficiency, which is one of the most serious problems in CPM systems.

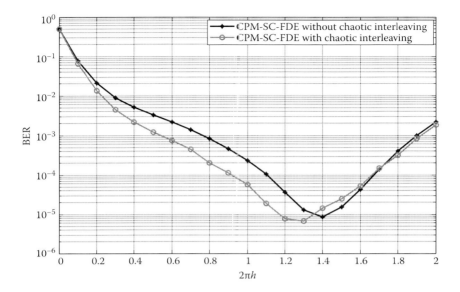

Figure 6.15 BER performance of the CPM-SC-FDE system with and without chaotic interleaving versus the modulation index at an SNR = 20 dB.

Table 6.2 Improvement Percentage (**G**) n Spectral Efficiency in the CPM-SC-FDE System with and without Chaotic Interleaving

BER	CPM-SC-FDE (NO INTERLEAVING)	CPM-SC-FDE (WITH CHAOTIC INTERLEAVING)	**G** (%)
1×10^{-4}	$\eta = 1.818$ bps $(2\pi h = 1.1)$	$\eta = 2$ bps $(2\pi h = 0.9)$	10
8×10^{-6}	$\eta = 1.43$ bps $(2\pi h = 1.4)$	$\eta = 1.67$ bps $(2\pi h = 1.2)$	16.8

Figure 6.16 shows the performance of the conventional SC-FDE system [9], the CPM-SC-FDE system without interleaving [18], and the proposed CPM-SC-FDE system with chaotic interleaving, in terms of the BER versus the RMS delay spread, which is normalized to the symbol period. The MMSE equalizer is used with all systems at a fixed SNR = 20 dB. As shown in this figure, in flat fading (i.e., $\tau_{rms} = 0$), the performance of the CPM-SC-FDE system converges with a small performance loss to the conventional OFDM system and the SC-FDE system, due to the effect of the phase demodulator threshold. In frequency-selective channels ($\tau_{rms} > 0$), however, the proposed CPM-SC-FDE system with chaotic interleaving achieves a significant performance gain over the other systems. It also provides a better performance than that in the flat fading case by exploiting the channel frequency diversity efficiently, especially at high delay spreads.

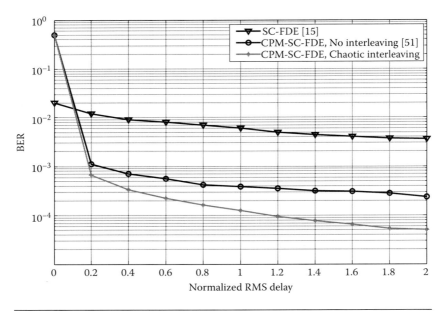

Figure 6.16 Effect of the RMS delay on the performance of the conventional SC-FDE system, the CPM-SC-FDE system, and the proposed CPM-SC-FDE system at an SNR = 20 dB.

6.7 Summary and Conclusions

In this chapter, an efficient chaotic interleaving scheme has been proposed for the CPM-OFDM and CPM-SC-FDE systems. The proposed scheme improves the BER performance of both systems more than the traditional block interleaving scheme, where it generates permuted sequences from the samples to be transmitted with lower correlation. The performance of the proposed CPM-OFDM system and CPM-SC-FDE system with chaotic interleaving was studied over a multipath fading channel with MMSE equalization. The obtained results show a noticeable performance improvement achieved by the proposed systems over the conventional CPM-OFDM system and the conventional CPM-SC-FDE without interleaving, especially at high RMS delay spreads. Simulation results have shown that the proposed CPM-OFDM system and the CPM-SC-FDE system with chaotic interleaving make a good trade-off between performance and spectral efficiency, where they achieve an efficient utilization of the frequency diversity and maintain the high spectral efficiency.

PART IV

7

Efficient Image Transmission over OFDM- and MC-CDMA-Based Systems

7.1 Introduction

This chapter proposes a new approach for efficient image transmission over systems based on multi-carrier orthogonal frequency division multiplexing (MC-OFDM) and multi-carrier code division multiple access (MC-CDMA) using chaotic interleaving. The chaotic interleaving scheme based on the Baker map is applied on the image data prior to transmission. The proposed approach transmits images over wireless channels efficiently without posing significant constraints on the wireless communication system bandwidth and noise. The performance of the proposed approach is further improved by applying frequency domain equalization (FDE) at receiver. Two types of frequency domain equalizers are considered and compared for performance evaluation of the proposed systems: the zero forcing (ZF) equalizer and the linear minimum mean square error (LMMSE) equalizer. Several experiments are carried out to test the performance of the image transmission with different sizes over the OFDM and MC-CDMA based systems. Simulation results show that image transmission over wireless channels using the proposed chaotic interleaving approach is much more immune to noise and fading. In addition, this chaotic interleaving process adds a degree of encryption to the transmitted data The results also show a noticeable performance improvement in terms of root-mean-square error (RMSE) and peak signal-to-noise ratio (PSNR) values when applying FDE to the proposed approach, especially LMMSE equalizer.

Reliable transmission of high-quality images through wireless channels represents a great challenge, especially for real-time applications. This is due to wireless channel impairments such as fading and interference which increase the probability of errors when the image data is transmitted [102]. The limited bandwidth restriction of wireless channel increases the demand for more reliable image communication systems that do not need higher bandwidths for achieving better image quality.

Multi-carrier modulation such as OFDM has a promising future as a new technology in many next-generation wireless communication systems [1]. In particular, combining OFDM and CDMA systems results in the MC-CDMA system [3,13,14]. The MC-CDMA system has received much attention among researchers. However, it suffers from multiple access interference (MAI) in a multiuser setting, which decreases the overall bit error rate (BER) performance. Multiuser detection (MUD) techniques have been introduced to mitigate MAI in order to improve system performance [15,16]. This chapter studies the efficient image transmission over MC-CDMA-based systems.

Transmission of images and multimedia with MC-CDMA technology has attracted the attention of several researchers [103,104]. In [103], a study of real-time image traffic over a radio link using wavelet transform (WT) to compress images and a CDMA link to transfer images over wireless communications network was presented. The transmission of compressed images through MC-CDMA channel in low SNR environment was considered in [104]. In addition, several coding techniques have been investigated for efficient transmission of images with MC-CDMA over wireless channels [103,105,106]. Despite their efficiency, these coding techniques add much redundancy to the transmitted data, which increases the bandwidth and reduces channel utilization.

Interleaving techniques are required to transmit data efficiently over wireless channels and combat problems caused by the multipath fading environment. There are several primary interleaving techniques such as block interleaving and helical interleaving [91,92,107]. In spite of the success of these techniques in achieving a good performance in wireless communication systems, there is a need for a much more powerful technique for severe channel degradation in image transmission. Chaotic Baker maps have been proposed for a wide range of applications in communications [93] and cryptography [94–97]. Due to the inherent strong randomization ability of these maps, they can be efficiently used for data interleaving. In References [98–101], a chaotic interleaving scheme has been proposed to improve the performance of the continuous phase modulation–based OFDM (CPM-OFDM) and continuous phase modulation–based single-carrier frequency domain equalization (CPM-SC-FDE) systems, respectively.

The main aim of this chapter is to propose a new approach for efficient image transmission over MC-CDMA-based systems using the chaotic interleaving technique proposed in [101]. Chaotic interleaving is applied on the image data prior to the modulation step in the transmitter side. The proposed approach transmits images over wireless channels efficiently, without posing significant constraints on the wireless communication system bandwidth and noise. With chaotic interleaving, any loss of an adjacent group of samples will affect different symbols rather than being concentrated on a single symbol as in the case without interleaving. The performance of the proposed approach is further improved by applying FDE at the receiver. Two types of frequency domain equalizers are considered: the ZF equalizer and the MMSE equalizer.

The rest of this chapter is organized as follows. Section 7.2 presents the proposed MC-CDMA system model. The proposed chaotic interleaving scheme is explained in Section 7.3. Linear equalization is explained in Section 7.4. Section 7.5 presents the simulation results and discussion. Finally, Section 7.6 gives concluding remarks.

7.2 OFDM System Model for Image Transmission

The proposed OFDM system model for image transmission consists mainly of four stages: image data formatting stage, chaotic interleaving stage on the binary image data (as explained in Chapter 6), OFDM modulation stage, and FDE stage at the receiver, as shown in Figure 7.1. OFDM modulator and demodulator are explained in Chapter 2.

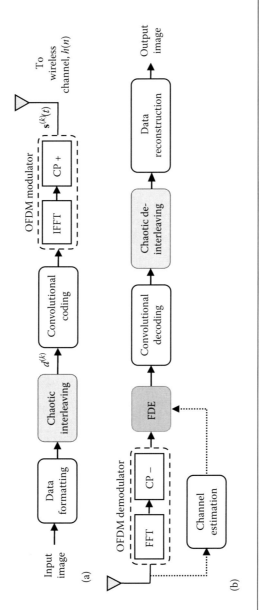

Figure 7.1 The OFDM system model for image transmission. (a) Transmitter and (b) receiver.

7.3 MC-CDMA System Model for Image Transmission

The basic MC-CDMA signal is generated by a serial concatenation of classical direct sequence CDMA (DS-CDMA) and OFDM. Each chip of the direct sequence spread data symbol is mapped onto a different subcarrier. Thus, with MC-CDMA, the chips of a spread data symbol are transmitted in parallel on different subcarriers, in contrast to a serial transmission with DS-CDMA [36]. The proposed MC-CDMA system model for image transmission consists mainly of five stages: image data formatting stage, chaotic interleaving stage on the binary image data, spreading and scrambling stage, OFDM modulation stage, and FDE stage at the receiver, as shown in Figure 7.2. In the transmitter, the complex-valued data symbol $d^{(k)}$ assigned to the kth user is multiplied by the kth spreading code, $\mathbf{c}^{(k)}$ of length N [36,107]:

$$\mathbf{c}^{(k)} = \left(\mathbf{c}_0^{(k)}, \mathbf{c}_1^{(k)}, \ldots, \mathbf{c}_{N-1}^{(k)}\right)^T \tag{7.1}$$

The chip rate of the serial spreading code $\mathbf{c}^{(k)}$ before serial-to-parallel conversion is

$$\frac{1}{T_c} = \frac{N}{T_d} \tag{7.2}$$

It is clear that the chip rate is N times higher than the data symbol rate $1/T_d$. The complex-valued sequence obtained after spreading can be expressed in vector notation as

$$\mathbf{S}^{(k)} = d^{(k)}\mathbf{c}^{(k)} = \left(\mathbf{S}_0^{(k)}, \mathbf{S}_1^{(k)}, \ldots, \mathbf{S}_{N-1}^{(k)}\right)^T \tag{7.3}$$

The sequence $\mathbf{S}^{(k)}$ is applied to a serial-to-parallel converter (not shown in the figure) and then modulated by an OFDM modulator to get the transmitted sequence, $\mathbf{s}^{(k)}(t)$. The CP is inserted to reduce intersymbol interference (ISI). At the receiver side, the received vector $\mathbf{r}$ is given by [107]

$$\mathbf{r} = \mathbf{H}\mathbf{s} + \mathbf{n} = \left(R_0, R_1, \ldots, R_{N-1}\right)^T \tag{7.4}$$

where

$\mathbf{H}$ is the $N \times N$ channel matrix

$\mathbf{n}$ is the noise vector of length N

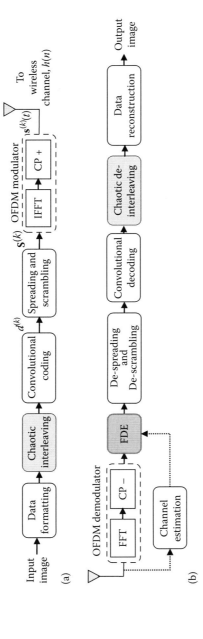

Figure 7.2   The MC-CDMA system model for image transmission. (a) Transmitter and (b) receiver.

The vector **r** is fed to the data detector in order to get a hard or soft estimation of the transmitted data.

7.4 Linear Equalization

Linear equalization is an efficient technique to suppress the ISI caused by the multipath environment and thereby improve the performance of the proposed MC-CDMA system. There are various types of linear equalization in the frequency domain, such as the LMMSE equalizer and the ZF equalizer [98–100].

- The ZF coefficients can be written as [101,108]

$$\mathbf{W}_{ZF} = \left(\mathbf{H}^{H}\mathbf{H}\right)^{-1}\mathbf{H}^{H} \tag{7.5}$$

  where **H** is the channel matrix. The drawbacks of the frequency domain ZF equalizer are noise enhancement and high time consumption in matrix inversion computations. However, its advantage is that the statistics of the additive noise and source data are not required.

  For the given statistics of the additive noise and the user's data, the equalizer can minimize the mean square error and partially remove the ISI. This equalizer is called the LMMSE equalizer. The LMMSE equalizer normally outperforms the ZF linear equalizer because of its better treatment of the noise.
- The LMMSE coefficients can be expressed as [101]

$$\mathbf{W}_{LMMSE} = \left(\mathbf{H}^{H}\mathbf{H} + \frac{1}{SNR}\mathbf{I}\right)^{-1}\mathbf{H}^{H} \tag{7.6}$$

The proposed MC-CDMA system model can be summarized in the following steps [107]:

1. The image is transformed to the binary format, pixel by pixel, in matrix form.
2. The nonsquare binary matrix is reshaped to an $M \times M$ square structure.
3. Chaotic randomization is applied to this square matrix.
4. After the randomization process, the square binary matrix is reshaped again to its original dimensions.

5. An MC-CDMA modulation step is performed on the binary data.
6. At the receiver, after the MC-CDMA demodulation and equalization steps, the received binary data is reshaped again to a square matrix and chaotic derandomization is applied to this matrix.
7. The square binary matrix after derandomization is reshaped again into its original dimensions.
8. Finally, the image is retrieved from the binary data.

7.5 Simulation Results and Discussion

7.5.1 Simulation Results for the OFDM System

In this section, several experiments are carried out to test the performance of the image communication system with OFDM. Two metrics are used to measure the quality of the reconstructed image compared to the original transmitted image. The first metric is the RMSE for the entire image, which is defined as [19,103,107]

$$\text{RMSE} = \frac{\sum_{i=1}^{M}\sum_{j=1}^{M}\left(I_0 - I_r\right)^2}{M^2} \tag{7.7}$$

The second metric is the PSNR, which is used to measure the quality of the reconstructed images at the receiver. PSNR is defined as the ratio between the maximum possible power of a signal and the power of corrupting noise that affects the fidelity of this signal. The PSNR can be expressed as follows [103,107]:

$$\text{PSNR} = 10\log_{10}\left(\frac{\text{max}_f^2}{\text{MSE}}\right) \tag{7.8}$$

where
M^2 is the total number of pixels in the images
I_0 and I_r are the values of pixels in the original and the recovered image

The 256 × 256 cameraman image shown in Figure 7.3 is used in the simulation experiments. Both additive white Gaussian noise (AWGN) and fading channels are considered in the simulation

Figure 7.3   The original cameraman image.

experiments. Channel equalization is used in some of the experiments. Convolutional coding is adopted as the error correction coding scheme in some of the experiments. Each user transmits binary phase shift keying (BPSK) information symbols. The wireless channel model used in the simulation experiments is the SUI-3 channel model [103], which is one of six channel models adopted by the IEEE 802.16a standard for evaluating the performance of broadband wireless systems in the 2–11 GHz band. It has three Rayleigh fading taps at delays of 0, 0.5, and 1 μs, with relative powers of 0, –5, and –10 dB, respectively. The fading is modeled as quasi-static (unchanging during a block). The simulation parameters are summarized in Table 7.1.

In the first experiment, the cameraman image is transmitted with OFDM. Several cases are considered and compared [107]:

a. No interleaving, coding, or equalization
b. Chaotic interleaving only
c. ZF equalization only

Table 7.1   Simulation Parameters

Transmitter	Modulation	BPSK
	Image size	256×256
	Cyclic prefix	20 samples
	Transmitter IFFT size	$M = 256$ symbols
	Chaotic map size	256×256
	Coding scheme	Convolutional
Channel	Fading channel	SUI-3 channel
	Noise environment	AWGN
Receiver	Equalization	ZF and LMMSE
	Channel estimation	Perfect

  d. Chaotic interleaving and ZF equalization
  e. LMMSE equalization
  f. Chaotic interleaving and LMMSE equalization
  g. Convolutional coding and ZF equalization
  h. Chaotic interleaving, convolutional coding, and ZF equalization
  i. Convolutional coding and LMMSE equalization
  j. Chaotic interleaving, convolutional coding, and LMMSE equalization

Figures 7.4 through 7.8 show the received images for all 10 cases at SNRs from 15 to 35 dB in 5 dB steps. Figure 7.9 shows a comparison between all cases. These figures reveal that the best PSNR results for image transmission with OFDM are obtained with chaotic interleaving, convolutional coding, and LMMSE equalization.

The PSNR values of the received cameraman image with OFDM system over the SUI-3 channel are given in Table 7.2.

7.5.2 Simulation Results for the MC-CDMA System

In this section, several experiments are carried out to test the performance of the image transmission over the proposed MC-CDMA system using the metrics discussed.

To evaluate the performance and efficiency of the proposed scheme, various monochrome grayscale images are used as input to the simulation framework, such as the cameraman image of size 256×256 shown in Figure 7.3 and the Lena image of size 128×128 shown in Figure 7.16. We consider the Lena image to check the performance of the MC-CDMA system with the proposed chaotic interleaving scheme when using a nonsquare binary matrix. As explained earlier, in this case we reshape it to $M \times M$ square structures and apply randomization by chaotic interleaving to each structure. The simulation environment is based on the MC-CDMA system, in which each user transmits BPSK information symbols. The wireless channel model used in the simulation is the SUI-3 channel, which is one of six channel models adopted by the IEEE 802.16a standard for evaluating the performance of broadband wireless systems in the 2–11 GHz band [22]. It has three Rayleigh fading taps at delays of 0, 0.5, and 1 s, with relative powers of 0, –5, and –10 dB, respectively. The fading is modeled as quasi-static (unchanging during a block). The simulation parameters are shown in Table 7.3.

Figure 7.4 The received cameraman image with OFDM in the 10 cases over an SUI-3 Rayleigh fading channel at SNR = 15 dB. (a) No interleaving, coding, or equalization. PSNR = 13.18 dB, (b) chaotic interleaving only. PSNR = 13.92 dB, (c) ZF equalization only. PSNR = 24.81 dB, (d) chaotic interleaving and ZF equalization. PSNR = 26.88 dB, (e) LMMSE equilibrium only. PSNR = 29.34 dB, (f) chaotic interleaving and LMMSE equalization. PSNR = 30.34 dB, (g) convolutional coding and ZF equalization. PSNR = 30.7 dB, (h) chaotic interleaving, convolutional coding, and ZF equalization. PSNR = 32 dB, (i) convolutional coding and LMMSE equalization. PSNR = 31.83 dB, and (j) chaotic interleaving, convolutional coding, and LMMSE equalization. PSNR = 32.74 dB.

Figure 7.5 The received cameraman image with OFDM in the 10 cases over an SUI-3 Rayleigh fading channel at SNR = 20 dB. (a) No interleaving, coding, or equalization. PSNR = 13.53 dB, (b) chaotic interleaving only. PSNR = 14.17 dB, (c) ZF equalization only. PSNR = 31.79 dB, (d) chaotic interleaving and ZF equalization. PSNR = 33.8 dB, (e) LMMSE equalization only. PSNR = 34.16 dB, (f) chaotic interleaving and LMMSE equalization. PSNR = 36.17 dB, (i) convolution and LMMSE equalization. PSNR = 42.56 dB, (j) chaotic interleaving, convolutional coding, and LMMSE equalization. PSNR = 45.04 dB.

Figure 7.6 The received cameraman image with OFDM in the 10 cases over an SUI-3 Rayleigh fading channel at SNR = 25 dB. (a) No interleaving, coding, or equalization. PSNR = 13.54 dB, (b) chaotic interleaving only. PSNR = 14.42 dB, (c) ZF equalization only. PSNR = 37.92 dB, (d) chaotic interleaving and ZF equalization. PSNR = 40.5 dB, (e) LMMSE equalization only. PSNR = 40.23 dB, (f) chaotic interleaving and LMMSE equalization. PSNR = 46.69 dB, (g) convolutional coding and ZF equalization. PSNR = 47.8 dB, (h) chaotic interleaving, convolutional coding, and ZF equalization. PSNR = 53.62 dB, (i) convolutional coding and LMMSE equalization. PSNR = 51.14 dB, (j) chaotic interleaving, convolutional coding, and LMMSE equalization. PSNR = 60. 04 dB.

Figure 7.7 The received cameraman image with OFDM in the 10 cases over an SUI-3 Rayleigh fading channel at SNR = 30 dB. (a) No interleaving, coding, or equalization. PSNR = 14.22 dB, (b) chaotic interleaving only. PSNR = 14.73 dB, (c) ZF equalization only. PSNR = 40.76 dB, (d) chaotic interleaving and ZF equalization. PSNR = 46.63 dB, (e) LMMSE equalization only. PSNR = 43.53 dB, (f) chaotic interleaving and LMMSE equalization. PSNR = 53.31 dB, (g) convolutional coding and ZF equalization. PSNR = inf., (h) chaotic interleaving, convolutional coding, and ZF equalization. PSNR = inf., (i) convolutional coding and LMMSE equalization. PSNR = inf., (j) chaotic interleaving, convolutional coding, and LMMSE equalization. PSNR = inf.

Figure 7.8 The received cameraman image with OFDM in the 10 cases over an SUI-3 Rayleigh fading channel at SNR = 35 dB. (a) No interleaving, coding, or equalization. PSNR = 14.27 dB, (b) chaotic interleaving only. PSNR = 14.91 dB, (c) ZF equalization only. PSNR = 45.12 dB, (d) chaotic interleaving and ZF equalization. PSNR = 54.06 dB, (e) LMMSE equalization only. PSNR = 48.12 dB, (f) chaotic interleaving and LMMSE equalization, (g) convolutional coding and ZF equalization. PSNR = inf., (h) chaotic interleaving, convolutional coding, and ZF equalization. SNR = 35; PSNR = inf., (i) convolutional coding and LMMSE equalization. PSNR = inf., (j) chaotic interleaving, convolutional coding, and LMMSE equalization. PSNR = inf.

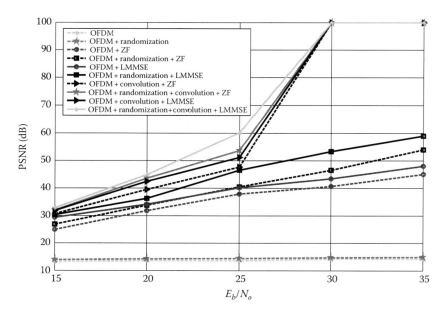

Figure 7.9 Variation of the PSNR of the received image with the SNR in the channel for image communication with OFDM over the SUI-3 channel in the 10 cases.

Table 7.2 Simulation Results of the Received Cameraman Image with OFDM

SNR	15 (dB)	20 (dB)	25 (dB)	30 (dB)	35 (dB)	40 (dB)
No interleaving, coding, or equalization	13.18	13.52	13.54	14.22	14.27	14.4
Chaotic interleaving only	13.92	14.17	14.41	14.72	14.91	15.02
ZF equalization only	24.81	31.79	37.91	42.75	45.11	48.13
Chaotic interleaving and ZF equalization	26.88	33.8	40.5	46.63	54.06	81.24
LMMSE equalization	29.34	34.15	40.23	43.53	48.11	54.08
Chaotic interleaving and LMMSE equalization	30.33	36.16	46.68	53.30	58.91	Inf.
Convolutional coding and ZF equalization	30.7	39.54	47.8	inf.	inf.	inf.
Chaotic interleaving, convolutional coding, and ZF equalization	32	43.67	53.62	inf.	inf.	inf.
Convolutional coding and LMMSE equalization	31.83	42.56	51.14	inf.	inf.	inf.
Chaotic interleaving, convolutional coding, and LMMSE equalization	32.74	45.04	60.04	inf.	inf.	inf.

Table 7.3 Simulation Parameters

Transmitter	Modulation	BPSK
	Spreading codes	Orthogonal variable spreading factor (OVSF) codes with processing gain 8
	Cyclic prefix	20 samples
	Transmitter IFFT size	256 symbols
	Chaotic map size	256 × 256 For cameraman image
		128 × 128 For Lena image
Channel	Fading	SUI-3 channel
	Noise environment	AWGN
Receiver	Equalization	ZF and LMMSE
	Channel estimation	Perfect

In the first experiment, the cameraman image of size 256 × 256 is transmitted with MC-CDMA. Several cases are considered and compared [107]:

a. Conventional MC-CDMA with no interleaving and no equalization
b. Chaotic interleaving only
c. ZF equalization only
d. Chaotic interleaving and ZF equalization
e. LMMSE equalization only
f. Chaotic interleaving and LMMSE equalization

Figures 7.10 through 7.13 show the received images for all six cases at SNRs from 15 to 30 dB in 5 dB steps. The obtained PSNR values are summarized in Table 7.4. Table 7.5 gives the RMSE value for each case.

Both the PSNR and RMSE values of the cameraman image of size 256 × 256 are given in Tables 7.4 and 7.5, respectively. From the tables, it is obvious that the quality of the reconstructed image degrades at a lower SNR environment. The higher the SNR value, the better the reconstructed image quality. It is also clear that the quality of the reconstructed image improves when moving from case (a), the conventional MC-CDMA system with no interleaving and no equalization to case (f), the proposed MC-CDMA system with chaotic interleaving and LMMSE equalization.

Figures 7.14 and 7.15 show the PSNR and RMSE values of the cameraman image, respectively, in terms of SNR. These results reveal

Figure 7.10 The received cameraman image with MC-CDMA using the six cases (from a to f) over SUI-3 Raleigh fading channel at SNR = 15 dB. (a) PSNR = 13.24 dB, (b) PSNR = 13.38 dB, (c) PSNR = 27.4 dB, (d) PSNR = 28.12 dB, (e) PSNR = 31.89 dB, and (f) PSNR = 32.86 dB.

Figure 7.11 The received cameraman image with MC-CDMA using the six cases (from a to f) over SUI-3 Raleigh fading channel at SNR = 20 dB. (a) PSNR = 13.34 dB, (b) PSNR = 13.47 dB, (c) PSNR = 33.43 dB, (d) PSNR = 35.68 dB, (e) PSNR = 37.47 dB, and (f) PSNR = 39.47 dB.

Figure 7.12 The received cameraman image with MC-CDMA using the six cases (from a to f) over SUI-3 Raleigh fading channel at SNR = 25 dB. (a) PSNR = 13.55 dB, (b) PSNR = 13.57 dB, (c) PSNR = 38.60 dB, (d) PSNR = 43.77 dB, (e) PSNR = 52.97 dB, and (f) PSNR = 59.90 dB.

Figure 7.13 The received cameraman image with MC-CDMA using the six cases (from a to f) over SUI-3 Raleigh fading channel at SNR = 30 dB. (a) PSNR = 15.22 dB, (b) PSNR = 15.79 dB, (c) PSNR = 45.49 dB, (d) PSNR = 51.13 dB, (e) PSNR = 84.25 dB, and (f) PSNR = inf.

Table 7.4 PSNR for the Cameraman Image in dB

CASES (a:f)	SNR (dB) 15	20	25	30
(a)	13.24	13.34	13.55	15.22
(b)	13.38	13.47	13.57	15.79
(c)	27.40	33.43	38.60	45.49
(d)	28.12	35.68	43.77	51.13
(e)	31.89	37.47	52.97	84.25
(f)	32.86	39.47	59.90	inf.

Table 7.5 RMSE for the Cameraman Image

CASES (a:f)	SNR (dB) 15	20	25	30
(a)	55.53	54.89	53.58	44.21
(b)	54.64	54	53.46	41.40
(c)	10.87	5.43	3	1.35
(d)	10	4.19	1.65	0.71
(e)	6.48	3.41	0.57	0.015
(f)	5.80	2.71	0.25	0

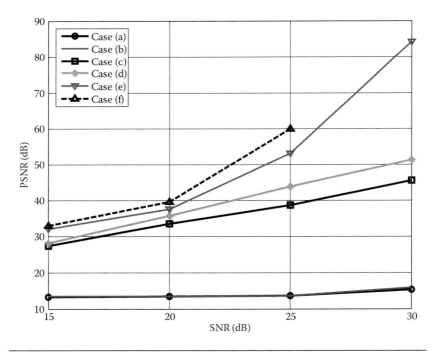

Figure 7.14 PSNR values for the cameraman image considering the six cases.

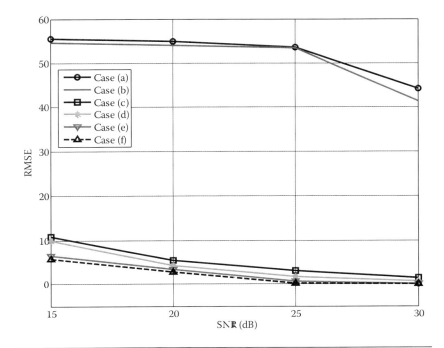

Figure 7.15 RMSE values for the cameraman image considering the six cases.

that the best PSNR and RMSE results for image transmission with MC-CDMA are obtained with chaotic interleaving and LMMSE equalization.

In the second experiment, Lena image of size 128 × 128 is transmitted with MC-CDMA to check the validity of our system to work with different image sizes. Several cases are considered and compared (Figure 7.16).

Figure 7.16 The original Lena image.

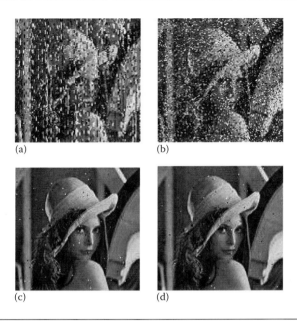

Figure 7.17   The received Lena image with MC-CDMA using the four cases (from a to d) over SUI-3 Raleigh fading channel at SNR = 15 dB. (a) PSNR = 13.60 dB, (b) PSNR = 18.81 dB, (c) PSNR = 31 dB, and (d) PSNR = 32.01 dB.

Figure 7.18   The received Lena image with MC-CDMA using the four cases (from a to d) over SUI-3 Raleigh fading channel at SNR = 20 dB. (a) PSNR = 13.79 dB, (b) PSNR = 13.90 dB, (c) PSNR = 37.63 dB, and (d) PSNR = 40.65 dB.

Figure 7.19 The received Lena image with MC-CDMA using the four cases (from a to d) over SUI-3 Raleigh fading channel at SNR = 25 dB. (a) PSNR = 13.80 dB, (b) PSNR = 13.97 dB, (c) PSNR = 47.85 dB, and (d) PSNR = 52.39 dB.

In this experiment, we check only the following cases for simplicity:

a. Conventional MC-CDMA with no interleaving and no equalization
b. Chaotic interleaving only
c. LMMSE equalization only
d. Chaotic interleaving and LMMSE equalization (Figures 7.17 through 7.19)

7.6 Summary and Conclusions

In this chapter, an efficient image transmission approach is proposed for both OFDM- and MC-CDMA-based systems using chaotic interleaving and frequency domain equalizers. The proposed approach transmits images over wireless channels efficiently, without posing significant constraints on the wireless communication system bandwidth and noise. The performance of the image transmission with different sizes was considered. The obtained results show a noticeable performance improvement in terms of RMSE and PSNR, especially when combining the chaotic interleaving process and the frequency domain equalization by LMMSE. In addition, the chaotic interleaving process adds a degree of encryption to the transmitted data.

8

EFFICIENT IMAGE TRANSMISSION OVER SC-FDMA-BASED SYSTEMS

8.1 Introduction

Present-day applications require various kinds of images as sources of information for interpretation and analysis. Single-carrier frequency division multiple access (SC-FDMA) is a promising technique for high data rate uplink communication. SC-FDMA is a recent variant of orthogonal frequency-division multiple access (OFDMA) in which the data symbols of each user are first modulated in the time domain and then DFT-spread across the data subcarriers. SC-FDMA generally exhibits a lower peak-to-average power (PAPR) because of its inherent single-carrier nature, and is therefore seen as an attractive alternative to OFDMA [1,2,117–119].

There are two variants of SC-FDMA which differ in the manner in which the subcarriers are mapped to a particular user. They are the interleaved FDMA (I-FDMA) [120], which assigns equidistant subcarriers to each user, and the localized FDMA (L-FDMA), which assigns groups of contiguous subcarriers to a particular user. With respect to immunity to transmission errors which determines throughput, I-FDMA is robust against frequency-selective fading because its information is spread across the entire signal band. Therefore, it offers the advantage of frequency diversity [121,122].

For high-efficiency image transmission, this chapter presents a comparison between two different systems—discrete Fourier transform (DFT)-based SC-FDMA and discrete cosine transform (DCT)-based SC-FDMA—in order to select the proper transmission technique for image transmission [20]. The obtained results show that the DCT-based SC-FDMA system achieves higher peak signal-to-noise ratio (PSNR) values than the DFT-based SC-FDMA system. The DCT has been implemented for both OFDM and SC-FDMA but the image communication problem has not been studied. In this chapter, we develop this approach to work with the SC-FDMA system and investigate its effect on image communication. This is because the DCT-based SC-FDMA system has excellent spectral energy compaction property, which makes most of the samples transmitted close to zero leading to a reduction in the effect of intersymbol interference (ISI). In addition, it uses basic arithmetic rather than the complex arithmetic used in the DFT. This reduces the signal processing complexity [123–125].

Hassan et al. [106–109] utilized chaotic interleaving to enhance the performance of the continuous phase modulation (CPM)-OFDM system for data transmission. As explained in Chapter 6, error bursts are better distributed to samples after de-interleaving in the proposed chaotic interleaving scheme than block interleaving mechanism. As a result, a better PSNR performance can be achieved when applying this scheme to the SC-FDMA system. Moreover, it adds a degree of security to the communication system.

This chapter studies image transmission over SC-FDMA-based systems [20]. The performance of two different structures—namely, DFT-based SC-FDMA and DCT-based SC-FDMA—will be studied in order to select the proper technique for efficient image transmission. We also use the chaotic interleaving scheme with both SC-FDMA structures for efficient image transmission. Both structures are simulated using MATLAB® and the experimental results show that the DCT-based SC-FDMA structure achieves higher PSNR values than the DFT-based SC-FDMA structure due to its excellent spectral energy compaction property. In addition, it uses basic arithmetic rather than the complex arithmetic used in the DFT-based SC-FDMA system. Furthermore, the results show that the PSNR values are enhanced by applying chaotic interleaving scheme in both structures.

We also study the performance of CPM-based DCT-SC-FDMA with image transmission [126]. The proposed structure combines the advantages of excellent spectral energy compaction property of DCT-based SC-FDMA, in addition to exploiting the channel frequency diversity and the power efficiency of CPM. Moreover, the performance of the proposed CPM-based DCT-SC-FDMA structure is compared with the conventional quadrature phase shift keying (QPSK)-based SC-FDMA system. Simulation experiments are performed using additive white Gaussian noise (AWGN) channel. Simulation results show that the CPM-based DCT-SC-FDMA structure increases the transmission efficiency and provides better performance and achieves higher PSNR values in the received images when compared to conventional QPSK-based SC-FDMA systems.

The rest of this chapter is organized as follows. In Section 8.2, image transmission over the SC-FDMA system is discussed. In Section 8.3, the QPSK-based DCT-SC-FDMA system is discussed. In Section 8.4, the DFT-based SC-FDMA architecture is explained. In Section 8.5, chaotic interleaving with SC-FDMA system is presented. Section 8.6 presents the proposed CPM-based SC-FDMA structure. In Section 8.7, the equalization is explained. In Section 8.8, the simulation results are presented.

8.2 Image Transmission over the SC-FDMA System

Digital images and videos have become an integral part of entertainment, business, and education in our daily lives. Studying image communication over evolving technologies such as SC-FDMA is therefore recommended. The image communication system with SC-FDMA is shown in Figure 8.1.

8.3 Conventional QPSK-Based DCT-SC-FDMA Structure

Figure 8.2 shows the structure of QPSK-based DCT-SC-FDMA. One disadvantage of the DFT for some applications is that the transform is complex valued, even for real data. The DCT does not have this problem. A DCT is a Fourier-related transform similar to the DFT, but using only real numbers. DCTs are equivalent to DFTs of

Figure 8.1   System model of image transmission over the SC-FDMA system.

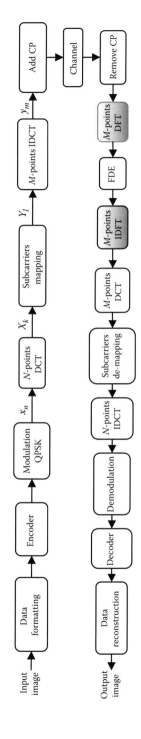

Figure 8.2 QPSK-based DCT- SC-FDMA structure using DFT and IDFT in the receiver [126].

roughly twice the length, operating on real data with even symmetry, where in some variants, the input and/or output data are shifted by half a sample. There are eight standard DCT variants, of which four are common [127].

Image formatting is used to transmit the image over the SC-FDMA system by converting it to a binary form suitable to be inserted and processed by the SC-FDMA System. The SC-FDMA transmitter starts with an encoder and then modulates the input signal using QPSK. Let x_n represent the modulated source symbols.

The signal after the DCT can be expressed as follows:

$$X_k = \sqrt{\frac{2}{N}} \beta_k \sum_{n=0}^{N-1} x_n \cos\left(\frac{\pi k(2n+1)}{2N}\right) \tag{8.1}$$

where x_n is the modulated data symbol, and β_k is given by

$$\beta_k = \begin{cases} \dfrac{1}{\sqrt{2}} & k = 0 \\ 1 & k = 1, 2, \ldots, N-1 \end{cases} \tag{8.2}$$

where N is the input block size, $\{x_n: n = 0, \ldots, N-1\}$ represents the modulated data symbols. The outputs are then mapped to M ($M > N$) orthogonal subcarriers followed by the M-points inverse DCT (IDCT). The subcarriers mapping assigns frequency domain modulation symbols to subcarriers Y_l, which represent the frequency domain sample after subcarriers mapping [126,128]. After IDCT, the signal can be expressed as follows:

$$y_m = \sqrt{\frac{2}{M}} \sum_{l=0}^{M-1} Y_l \beta_l \cos\left(\frac{\pi l(2m+1)}{2M}\right) \tag{8.3}$$

where

Y_l is the signal after subcarrier mapping

M is the IDCT length (number of subcarriers) ($M = Q \cdot N$)

Q is the bandwidth expansion factor of the symbol sequence

If all terminals transmit N symbols per block, the system can handle Q simultaneous transmissions without co-channel interference. We can also insert a set of symbols referred to as CP in order to provide a guard time to prevent interblock interference (IBI) due to multipath propagation. The CP is a copy of the last part of the block. It is inserted at the start of each block. The transmitted data propagates through the channel.

At the receiver, the CP is removed, and then the received signal is transformed into the frequency domain by fast Fourier transform (FFT) in order to recover subcarriers. The samples are passed through the frequency domain equalization (FDE). The inverse transform (IFFT) at the receiver transforms equalized symbols back to the time domain. After that, the time domain samples are passed through the DCT, then the de-mapping operation isolates the frequency domain samples of each source signal. The samples are then passed through the IDCT. The demodulation process recovers the original data, which is passed through the decoder. Image reconstruction is used to convert the binary form to an image to recover the original image. This causes complexity in the receiver.

8.4 DFT-Based SC-FDMA Structure

Image formatting is used to transmit the image over the SC-FDMA system by converting the image to a binary form suitable to be inserted and processed by the SC-FDMA system (Figure 8.3). The SC-FDMA transmitter starts the encoder and then modulates the input signal using binary QPSK. Let x_n represent the modulated source symbols. Then, the signal is transformed into frequency domain to produce frequency domain symbols X_K. The signal after the DFT can be expressed as follows:

$$X_K = \sum_{N=0}^{N-1} x_n e^{-\frac{j2\pi}{N}nk} \tag{8.4}$$

The outputs are then mapped to M ($M > N$) orthogonal subcarriers followed by the M-points IDFT to convert to a time domain complex signal sequence. $M = QN$ is the output block size. Q is the maximum number of users that can transmit. The subcarrier mapping assigns frequency domain modulation symbols to subcarrier Y_l which represents the frequency domain sample after subcarrier mapping [14].

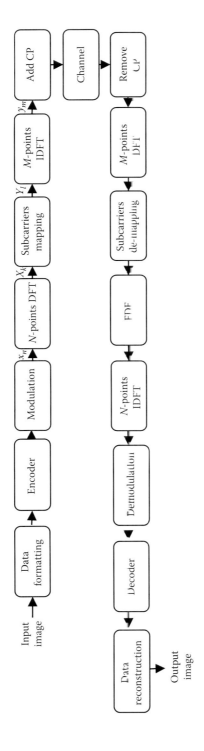

Figure 8.3 DFT-based SC-FDMA architecture.

The inverse transform creates a time domain symbol y_m. The resulting signal after the IDFT can be given as follows:

$$y_m = \frac{1}{M} \sum_{l=0}^{M-1} Y_l e^{j\frac{2\pi}{M}ml} \tag{8.5}$$

where $\{Y_l: l = 0, \ldots, M - 1\}$ represents the frequency domain samples after the subcarriers mapping scheme. It also inserts a set of symbols referred to as CP in order to provide a guard time to prevent IBI due to multipath propagation. The cyclic prefix is a copy of the last part of the block. It is inserted at the start of each block. The transmitted data propagates through the channel.

At the receiver, the CP is removed and then the received signal is transformed into the frequency domain in order to recover subcarriers. The de-mapping operation isolates the frequency domain samples of each source signal because SC-FDMA uses single-carrier modulation. The samples are then passed through the FDE. Then the inverse transform in the receiver transforms equalized symbols back to the time domain. The demodulation process recovers the original data, which is then passed through the decoder. Image reconstruction is used to convert the binary form to recover the original image.

8.5 SC-FDMA System with Chaotic Interleaving

The signal samples can be arranged into a 2-D format then randomized using the chaotic Baker map as explained in detail in Chapter 5. Figures 8.4 and 8.5 show the modified SC-FDMA system with chaotic interleaving.

8.6 CPM-Based SC-FDMA Structures

8.6.1 CPM-Based DFT-SC-FDMA Structure

Figure 8.6 shows the CPM-based DFT-SC-FDMA structure which is used in the performance comparison with the proposed structure.

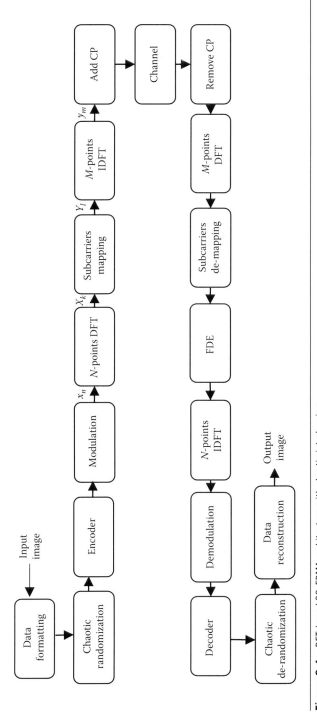

Figure 8.4 DFT-based SC-FDMA architecture with chaotic interleaving.

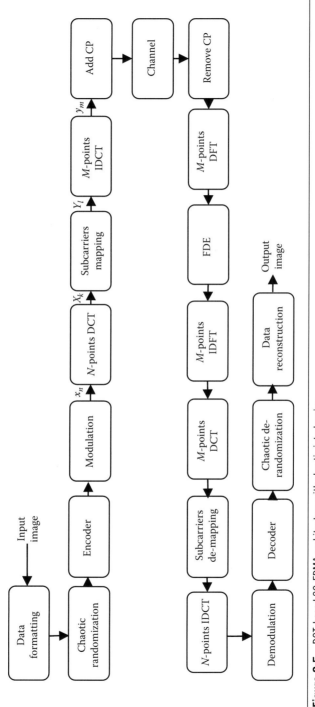

Figure 8.5 DCT-based SC-FDMA architecture with chaotic interleaving.

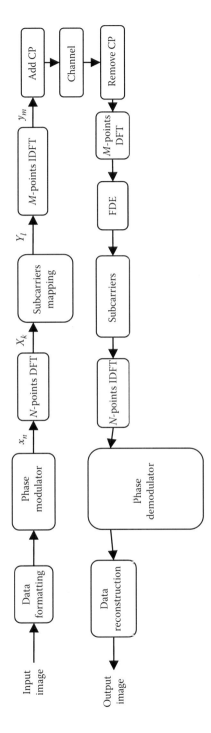

Figure 8.6 CPM-based DFT-SC-FDMA structure.

8.6.2 CPM-Based DCT-SC-FDMA Structure

See Figure 8.7.

8.7 Equalizer Design

The FDE cancels the ISI. The equalizer coefficients $W(m)$ are computed according to the type of the FDE as follows:

- The zero forcing (ZF) equalizer:

$$W(m) = \frac{1}{H(m)} \quad m = 0, 1, \ldots, N_{DFT} - 1 \qquad (8.6)$$

- The minimum mean square error (MMSE) equalizer:

$$W(m) = \frac{H^*(m)}{|H(m)|^2 + (E_b/N_0)^{-1}} \qquad (8.7)$$

where $H(m)$ is the channel transfer function and $H^*(m)$ denotes the complex conjugate and (E_b/N_0) is the SNR. The ZF FDE perfectly reverses the effect of the channel in the absence of noise. In the presence of noise, it suffers from the noise enhancement phenomenon. On the other hand, the MMSE equalizer takes into account the SNR, making an optimum trade-off between channel inversion and noise enhancement [109].

8.8 Simulation Results and Discussions

The PSNR is used to measure the quality of the reconstructed images at the receiver. To evaluate the performance and efficiency of both systems, various monochrome images are used, such as the cameraman image (256 × 256, grayscale), which is used as an input to the simulation framework. The simulation parameters are as shown in Table 8.1. The simulation results are shown in Figures 8.8 through 8.10, and Figure 8.11 shows the PSNR values for the DFT-SC-FDMA and DCT-SC-FDMA systems without randomization.

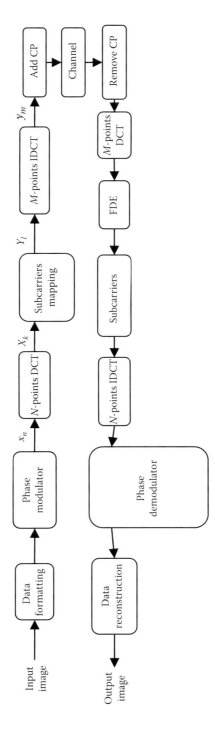

Figure 8.7 CPM-DCT SC-FDMA architecture without using DFT and IDFT in the receiver.

Table 8.1 Simulation Parameters

SIMULATION PARAMETER	VALUE
FFT size	512 symbols
Input block size	128 symbols
Cyclic prefix size	20 samples
Image size	256×256 for cameraman
Channel coding	Convolutional code with rate 1/2
Modulation type	QPSK
Subcarriers mapping	Interleaved and localized
Channel model	SUI3 channel
Noise environment	AWGN
Equalizer type	MMSE equalizer

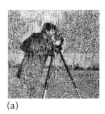 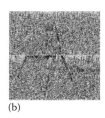  

(a)   (b)   (c)   (d)

Figure 8.8 Simulation results for SNR = 5 dB without randomization: (a) DCT-SC-FDMA(iFDMA), PSNR = 10.8446 dB; (b) DCT-SC-FDMA(IFDMA), PSNR = 8.8034 dB; (c) DFT-SC-FDMA(iFDMA), PSNR = 8.4958 dB; and (d) DFT-SC-FDMA(IFDMA), PSNR = 8.5382 dB.

  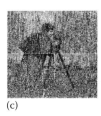 

(a)   (b)   (c)   (d)

Figure 8.9 Simulation results for SNR = 10 dB without randomization: (a) DCT-SC-FDMA(iFDMA), PSNR = 23.5637 dB; (b) DCT-SC-FDMA(IFDMA), PSNR = 15.2518 dB; (c) DFT-SC-FDMA(iFDMA), PSNR = 9.7824 dB; and (d) DFT-SC-FDMA(IFDMA), PSNR = 11.5079 dB.

(a)   (b)   (c)   (d)

Figure 8.10 Simulation results for SNR = 15 dB without randomization: (a) DCT-SC-FDMA(iFDMA), PSNR = 47.4384 dB; (b) DCT-SC-FDMA(IFDMA), PSNR = 35.1902 dB; (c) DFT-SC-FDMA(iFDMA), PSNR = 20.7386 dB; and (d) DFT-SC-FDMA(IFDMA) PSNR = 32.2037 dB.

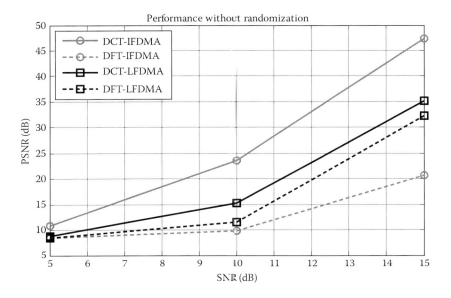

Figure 8.11   PSNR values for DFT-SC-FDMA and DCT-SC-FDMA systems without randomization.

Figures 8.8 through 8.10 show the simulation results when transmitting the original image through the DFT-SC-FDMA and DCT-SC-FDMA systems at different SNR values without chaotic interleaving (without randomization). The results show that the DCT-SC-FDMA system has higher PSNR values than the DFT-SC-FDMA system. This is because the DCT-based SC-FDMA system has excellent spectral energy compaction property, which makes most of the samples transmitted close to zero leading to a reduction in the effect of ISI. In addition, it uses basic arithmetic rather than the complex arithmetic used in DFT. This reduces the signal processing complexity.

Figures 8.12 through 8.16 show the simulation results when applying chaotic interleaving. The results show that the DCT-SC-FDMA system with the chaotic interleaving has higher PSNR values than DCT-SC-FDMA system, where error bursts are better distributed to samples after de-interleaving in the proposed chaotic interleaving scheme. As a result, a better PSNR performance can be achieved when applying this scheme. It also adds a degree of security to the communication system.

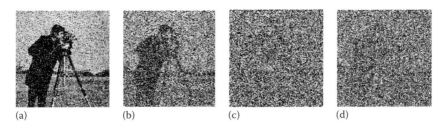

(a)   (b)   (c)   (d)

Figure 8.12 Simulation results for SNR = 5 dB with randomization: (a) DCT-SC-FDMA(iFDMA), PSNR = 13.9766 dB; (b) DCT-SC-FDMA(IFDMA), PSNR = 9.8538 dB; (c) DFT-SC-FDMA(iFDMA), PSNR = 8.5848 dB; and (d) DFT-SC-FDMA(IFDMA), PSNR = 8.8137 dB.

(a)   (b)   (c)   (d)

Figure 8.13 Simulation results for SNR = 10 dB with randomization: (a) DCT-SC-FDMA(iFDMA), PSNR = 31.6008 dB; (b) DCT-SC-FDMA(IFDMA), PSNR = 18.9805 dB; (c) DFT-SC-FDMA(iFDMA), PSNR = 10.6970 dB; and (d) DFT-SC-FDMA(IFDMA), PSNR = 15.6505 dB.

(a)   (b)   (c)   (d)

Figure 8.14 Simulation results for SNR = 15 dB with randomization: (a) DCT-SC-FDMA(iFDMA), PSNR = 69.2029 dB; (b) DCT-SC-FDMA(IFDMA), PSNR = 37.6353 dB; (c) DFT-SC-FDMA(iFDMA), PSNR = 22.0001 dB; and (d) DFT-SC-FDMA(IFDMA), PSNR = 39.1386 dB.

To evaluate the performance and efficiency of CPM-based DCT-SC-FDMA structures, the 256 × 256 cameraman image is used. The simulation parameters are given in Table 8.2 and the simulation results are shown in Figures 8.17 through 8.19. Figures 8.20 through 8.23 show a comparison between both schemes.

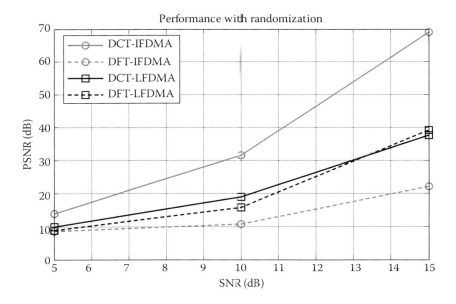

Figure 8.15   PSNR values for DFT-SC-FDMA and DCT-SC-FDMA systems with randomization.

Figures 8.17 through 8.23 show the simulation results when transmitting the original image through the QPSK-DCT-SC-FDMA, CPM-based DCT-SC-FDMA, and CPM-based DFT-SC-FDMA systems at different SNR values and at AWGN channel. The results show that the CPM-based DCT-SC-FDMA system achieves higher PSNR values than the QPSK-DCT-SC-FDMA system and that the CPM based DCT-SC-FDMA system achieves higher PSNR values than the CPM-based DFT-SC-FDMA system due to the energy efficiency of CPM and the energy compaction property of the DCT-based SC-FDMA system, which makes most of the samples transmitted close to zero, leading to a reduction in the effect ISI.

Table 8.3 summarizes the obtained results of the three different structures considered in this chapter. It is clear that the CPM-based DCT-SC-FDMA structure outperforms both the QPSK-based DCT-SC-FDMA and CPM-based DFT-SC-FDMA structures.

8.9 Summary and Conclusions

This chapter studied the efficient image transmission over the SC-FDMA system. Two different structures were used: DFT-SC-FDMA and DCT-SC-FDMA. Also, a chaotic interleaving scheme was applied to both structures for efficient image transmission. The experimental results have shown that the DCT-based SC-FDMA system has higher PSNR values than the DFT-based SC-FDMA system. This is because DCT-based SC-FDMA system has excellent spectral energy compaction property, which makes most of the samples transmitted close to zero leading to

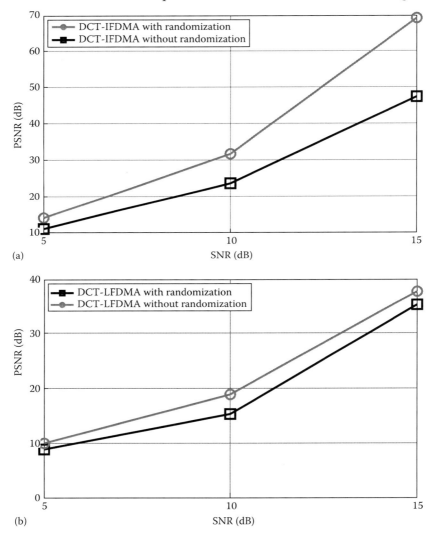

(a)

(b)

Figure 8.16   A comparison between the DCT-based SC-FDMA and DFT-based SC-FDMA systems with and without randomization by chaotic interleaving. (a) PSNR of DCT-IFDMA with and without randomization by chaotic interleaving. (b) PSNR of DCT-LFDMA with and without randomization by chaotic interleaving. *(Continued)*

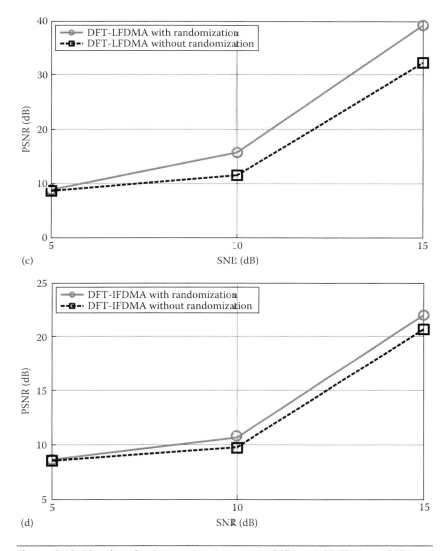

Figure 8.16 (*Continued*) A comparison between the DCT-based SC-FDMA and DFT-based SC-FDMA systems with and without randomization by chaotic interleaving. (c) PSNR of DFT-LFDMA with and without randomization by chaotic interleaving. (d) PSNR of DFT-IFDMA with and without randomization by chaotic interleaving.

a reduction in the effect of ISI. In addition, it uses basic arithmetic rather than the complex arithmetic used in the DFT. This reduces the signal processing complexity. The results also show a noticeable improvement in PSNR value when applying the chaotic interleaving scheme.

Table 8.2   Simulation Parameters

SIMULATION PARAMETER	VALUE
FFT size	512 symbols
Input block size	128 symbols
Cyclic prefix size	20 samples
Image size	256 × 256 for cameraman
Channel coding	Convolutional code with rate 1/2
Modulation type	QPSK and CPM
Subcarriers mapping	Interleaved and localized
Channel model	AWGN
Noise environment	AWGN
Equalizer type	MMSE equalizer

(a)   (b)   (c)   (d)

(e)   (f)

Figure 8.17   Simulation results for the QPSK-based DCT-SC-FDMA structure with AWGN channel. (a) PSNR-IFDMA = 23.4504 dB. SNR = 5, (b) PSNR-LFDMA = 23.5216 dB. SNR = 5, (c) PSNR-IFDMA = 34.3422 dB. SNR = 7, (d) PSNR-LFDMA = 34.8901 dB. SNR = 7, (e) PSNR-IFDMA = 58.3991 dB. SNR = 10, and (f) PSNR-LFDMA = 61.4101 dB. SNR = 10.

This chapter also presented an efficient image transmission scheme over the SC-FDMA system using CPM. Three different structures were considered: QPSK-based DCT-SC-FDMA, CPM-based DCT-SC-FDMA, and CPM-based DFT-SC-FDMA. Each structure was simulated using MATLAB to evaluate the performance and the efficiency of each structure. The experimental results showed that the CPM-based DCT-SC-FDMA structure achieves higher PSNR values than both the QPSK-based

Figure 8.18 Simulation results for the CPM-based DFT-SC-FDMA structure with AWGN channel. (a) PSNR-IFDMA = 17.9834 dB. SNR = 5, (b) PSNR-LFDMA = 17.8571 dB. SNR = 5, (c) PSNR-IFDMA = 21.0154 dB. SNR = 7, (d) PSNR-LFDMA = 20.7965 dB. SNR = 7, (e) PSNR-IFDMA = 28.2613 dB. SNR = 10, and (f) PSNR-LFDMA = 28.0819 dB. SNR = 10.

Figure 8.19 Simulation results for the CPM-based DCT-SC-FDMA structure with AWGN channel. (a) PSNR-IFDMA = 31.7764 dB. SNR = 5, (b) PSNR-LFDMA = 31.7672 dB. SNR = 5, (c) PSNR-IFDMA = 42.4094 dB. SNR = 7, (d) PSNR-LFDMA = 42.2956 dB. SNR = 7, (e) PSNR-IFDMA = Inf. SNR = 10, and (f) PSNR-LFDMA = Inf. SNR = 10.

DCT-SC-FDMA and CPM-based DCT-SC-FDMA structures. This is due to the energy efficiency of CPM-based systems and the excellent spectral energy compaction property of DCT-based systems, which makes most of the samples transmitted close to zero leading to a reduction in the effect of ISI. In addition, it uses basic

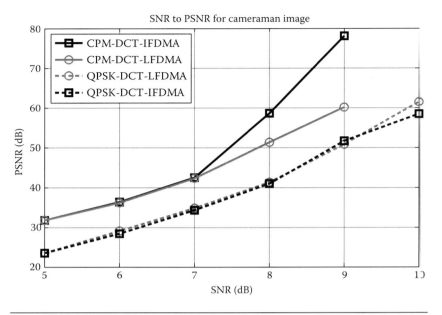

Figure 8.20 PSNR versus SNR for the CPM-DCT-SC-FDMA and QPSK-DCT-SC-FDMA systems at the AWGN channel.

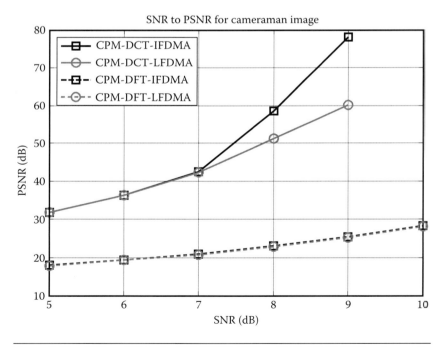

Figure 8.21 PSNR vs. SNR for the CPM-DCT-SC-FDMA and CPM-DFT-SC-FDMA systems at the AWGN channel.

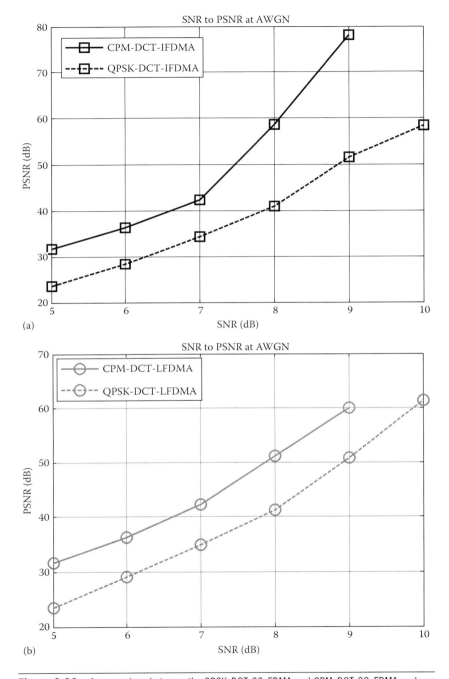

Figure 8.22 A comparison between the QPSK-DCT-SC-FDMA and CPM-DCT-SC-FDMA systems at the AWGN channel. (a) PSNR values of QPSK-DCT-IFDMA and CPM-DCT-IFDMA systems. (b) PSNR values of QPSK-DCT-LFDMA and CPM-DCT-LFDMA systems.

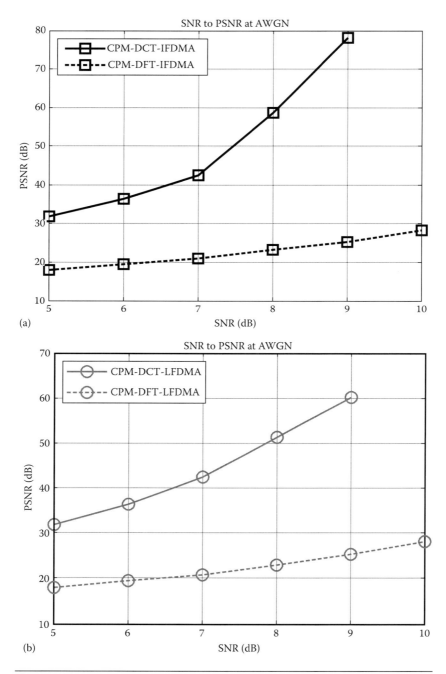

Figure 8.23 A comparison between the CPM-DCT-SC-FDMA and CPM-DFT-SC-FDMA systems at the AWGN channel.

Table 8.3 Simulation Results Summary

SNR	QPSK-BASED DCT-SC-FDMA		CPM-BASED DFT-SC-FDMA		CPM-BASED DCT-SC-FDMA	
	IFDMA	LFDMA	IFDMA	LFDMA	IFDMA	LFDMA
5	23.4504	23.5216	17.9834	17.8571	31.7764	31.7672
7	34.3422	34.8901	21.0154	20.7965	42.4094	42.2956
10	58.3991	61.4101	28.2613	28.0819	Inf	Inf

arithmetic rather than the complex arithmetic used in the DFT. This reduces the signal processing complexity. Also avoiding the DFT and IDFT at the receiver in CPM-based DCT-SC-FDMA reduces the receiver complexity in the uplink unlike the QPSK-based DCT-SC-FDMA system.

Appendix A: MATLAB® Simulation Codes for Chapters 1 through 4

```
%*******************************************************
% HPA transfer function
%*******************************************************

% Initialization
    clear all;
    close all;
    clc;

%=======================================================
                % Setting Parameters
%=======================================================
    % OFDM System Parameters
    N           = 256;      % length of OFDM IFFT
(16,32,64,...,2^n)
    M           = 16;       % number of QAM
constellation points (4,16,64,256)
    numOfZeros  = N/4+1;    % numOfZeros must be an
odd number and lower
                           % than N. The zero padding
operation is
                           % necessary in practical
implementations.
```

```
    GI          = 1/4;      % Guard Interval
(1/4,1/8,1/16,...,4/N)
    BW          = 20;       % OFDM signal Band width
in MHz
    numOfSym    = 100;      % number of OFDM Symbols

    % Amplifier Parameters
    satLevel = 5;    % in dB , higher than the tx out
mean of voltage

%========================================================
                    %    Main Program
%========================================================
    txData      = randint(N-numOfZeros,numOfSym,M);
% data generation

    % QAM modulation
    txDataMod   = qammod(txData,M);

    % zeros padding
    txDataZpad = [txDataMod((N-numOfZeros+1)/2:
end,:);...
                 zeros(numOfZeros,numOfSym);...
                 txDataMod(1:(N-numOfZeros+1)/2+1,:)];

    % IFFT
    txDataZpadIfft = sqrt(N)*ifft(txDataZpad,N);

    % Guard Interval Insertion
    txDataZpadIfftGI    = [txDataZpadIfft((1-
GI)*N+1:end,:);txDataZpadIfft];

    % Amplifier Model
    txDataZpadIfftGIAbs     = abs(txDataZpadIfftGI);
% tx data amplitude

    % tx data amplitude standard deviation and mean
    txDataZpadIfftGIAbsStd  = mean(std(txDataZpadIfftG
IAbs));
    txDataZpadIfftGIAbsMean = mean(mean(txDataZpadIfft
GIAbs));

    % tx data phase in radian
    txDataZpadIfftGIAng     = angle(txDataZpadIfftGI);
```

```
    txDataZpadIfftGIAbsHPA = txDataZpadIfftGIAbs ./...
sqrt(1+(txDataZpadIfftGIAbs/(txDataZpadIfftGIAbsMean
*10^(satLevel/10))).^2);
    % no change in the phase
    txDataZpadIfftGIAngHPA = txDataZpadIfftGIAng;

    % mean of amplitude after amplification
    txDataZpadIfftGIAbsHPAmean =
mean(mean(txDataZpadIfftGIAbsHPA));
    % standard deviation after amplification
    txDataZpadIfftGIAbsHPAStd  =
mean(std(txDataZpadIfftGIAbsHPA));

    % polar to Cartesian conversion
    txDataZpadIfftGIHPA    = txDataZpadIfftGIAbsHPA.*
...
                                exp(sqrt(-1) *
txDataZpadIfftGIAngHPA);

    % receiver part
    % Guard Interval removal
    rxDataZpadIfftHPA  = txDataZpadIfftGIHPA
(GI*N+1 : N+GI*N,:);
    % FFT operation
    rxDataZpadHPA  = 1/sqrt(N)*fft(rxDataZpadIfftHPA,N);
    % zero removal and rearrangement
    rxDataModHPA       = [rxDataZpadHPA((N-(N-
numOfZeros-1)/2+1):N,:);...
                         rxDataZpadHPA(1:(N-
numOfZeros+1)/2,:)];
    % demodulation
    rxDataHPA          = qamdemod(rxDataModHPA/mean
(std(rxDataModHPA))*mean(std(txDataMod)),M);

%========================================================
%       statistical computation
%========================================================
    % Mean Error Rate computation
    MER       = 10*log10(mean(var(rxDataModHPA./
mean(std(rxDataModHPA))...
                - txDataMod./mean(std(txDataMod)))));
    % Bit Error Rate computation
    [num BER] = symerr(rxDataHPA,txData);
%========================================================
```

```
%                % normalize amplitude
            txAmp = linspace(0,5,100);
            amAmp = txAmp./(1+(txAmp/satLevel).^2);
            plot(txAmp,txAmp,'b');
            hold on
            plot(txAmp,amAmp,'r');
            xlabel('Input Power');
            ylabel('Output Power');
            %title('AM/AM response of power amplifier');
            legend('Linear Response','Amplifier
Response');


%*******************************************************
%%% this program for SISO_SLM system: CCDF of PAPR of
an OFDM  signal
%*******************************************************

clear all
K=64;% No of subcarriers
alph=1;   % oversampling factor
K1=K*alph;
V=1;   % no. of phase sequences (V = 1, no SLM)
i=0;
for R=1:0.5:15
    i=i+1;
    P(i)=(1-(1-exp(-R))^K1)^V;
end
R=1:0.5:15;
Rd=10*log10(R);
semilogy(Rd,P,'b')
axis([5 12 1e-4 1e0])
grid on
hold on
xlabel('PAPRo [dB]')
ylabel('Prob (PAPRlow > PAPRo )')


%*******************************************************
% Effects of nonlinear PA on signal spectrum
%*******************************************************

% Initialization
    clear all;
    close all;
    clc;
```

```
%====================================================
                  % Setting Parameters
%====================================================
    % OFDM System Parameters
    N = 256;        % length of OFDM IFFT
(16,32,64,...,2^n)
    M = 16;         % number of QAM constellation points
(4,16,64,256)
    numOfZeros = N/4+1;    % numOfZeros must be an odd
number and lower than N.
%The zero padding operation is necessary in practical
%implementations.
    GI = 1/4;       % Guard Interval
(1/4,1/8,1/16,...,4/N)
    BW = 20;        % OFDM signal Band width in MHz
    numOfSym = 100;    % number of OFDM Symbols

    % Amplifier Parameters
    satLevel = 5;   % in dB , higher than the tx out
mean of voltage


%====================================================
                  %    Main Program
%====================================================
txData          = randint(N-numOfZeros,numOfSym,M);
% data generation

    % QAM modulation
    txDataMod  = qammod(txData,M);

    % zeros padding
    txDataZpad = [txDataMod((N-numOfZeros+1)/2:
end,:);...
                  zeros(numOfZeros,numOfSym);...
                  txDataMod(1:(N-numOfZeros+1)/2+1,:)];

    % IFFT
    txDataZpadIfft = sqrt(N)*ifft(txDataZpad,N);

    % Guard Interval Insertion
    txDataZpadIfftGI    = [txDataZpadIfft((1-
GI)*N+1:end,:);txDataZpadIfft];

    % Amplifier Model
```

```
    txDataZpadIfftGIAbs      = abs(txDataZpadIfftGI);
% tx data amplitude

    % tx data amplitude standard deviation and mean
    txDataZpadIfftGIAbsStd  = mean(std(txDataZpadIfftG
IAbs));
    txDataZpadIfftGIAbsMean = mean(mean(txDataZpadIfft
GIAbs));

    % tx data phase in radian
    txDataZpadIfftGIAng      = angle(txDataZpadIfftGI);

    txDataZpadIfftGIAbsHPA = txDataZpadIfftGIAbs ./...
sqrt(1+(txDataZpadIfftGIAbs/(txDataZpadIfftGIAbsMean
*10^(satLevel/10))).^2);
    % no change in the phase
    txDataZpadIfftGIAngHPA = txDataZpadIfftGIAng;

    % mean of amplitude after amplification
    txDataZpadIfftGIAbsHPAmean =
mean(mean(txDataZpadIfftGIAbsHPA));
    % standard deviation after amplification
    txDataZpadIfftGIAbsHPAStd  =
mean(std(txDataZpadIfftGIAbsHPA));


    % polar to Cartesian conversion
    txDataZpadIfftGIHPA      = txDataZpadIfftGIAbsHPA.
*  ...
                                exp(sqrt(-1) *
txDataZpadIfftGIAngHPA);

    % receiver part
    % Guard Interval removal
    rxDataZpadIfftHPA  = txDataZpadIfftGIHPA(GI*N+1 :
N+GI*N,:);
    % FFT operation
    rxDataZpadHPA       = 1/sqrt(N)*fft
(rxDataZpadIfftHPA,N);
    % zero removal and rearrangement
    rxDataModHPA        = [rxDataZpadHPA((N-(N-
numOfZeros-1)/2+1):N,:);...
                            rxDataZpadHPA(1:(N-
numOfZeros+1)/2,:)];
```

```
    % demodulation
    rxDataHPA          = qamdemod(rxDataModHPA/mean(std
(rxDataModHPA))*mean(std(txDataMod)),M);


%========================================================
%        statistical computation
%========================================================
    % Mean Error Rate computation
    MER       = 10*log10(mean(var(rxDataModHPA./
mean(std(rxDataModHPA))...
              - txDataMod./mean(std(txDataMod)))));
    % Bit Error Rate computation
    [num BER] = symerr(rxDataHPA,txData);
%========================================================

  f1 = figure(1);
  subplot(1,2,1)
  set(f1,'color',[1 1 1]);


          spectrumFftSize = 2*N;
          % spectrum of signal before High Power
Amplifier
          txSpec  =
20*log10(mean(abs(fft(txDataZpadIfftGI(:,:)./ ...
mean(std(txDataZpadIfftGI)),spectrumFftSize)),2));
          % spectrum of signal after High Power
Amplifier
          HpaSpec =
20*log10(mean(abs(fft(txDataZpadIfftGIHPA(:,:)./ ...
mean(std(txDataZpadIfftGIHPA)),spectrumFftSize)),2));
          % corresponding frequency
          Freq    =
linspace(-BW/2,BW/2,length(txSpec));

plot(Freq,[txSpec(length(txSpec)/2:length
(txSpec));...
                txSpec(1:(length(txSpec)/2-1))]);
          hold on
          grid on
plot(Freq,[HpaSpec(length(txSpec)/2:length
(txSpec));...
                HpaSpec(1:(length(txSpec)/2-1))],'r');
          grid on;
          xlabel('Frequency [MHz]');
```

```
            ylabel('Spectrum');
            title('Spectrum Effects')
            legend('Before Amplifier','After Amplifier')

 subplot(1,2,2)
            plot(real(reshape(rxDataModHPA,1,numOfSym
*(N-numOfZeros)))/mean(std(rxDataModHPA)),...
                imag(reshape(rxDataModHPA,1,numOfSym
*(N-numOfZeros)))/mean(std(rxDataModHPA)),'.r')
            hold on
            plot(real(reshape(txDataMod,1,numOfSym
*(N-numOfZeros)))/mean(std(txDataMod)),...
                imag(reshape(txDataMod,1,numOfSym
*(N-numOfZeros)))/mean(std(txDataMod)),'.b')
            xlabel('I channel');
            ylabel('Q channel');
            title('Signal Constellations');
            legend('After Amplifier','Before
Amplifier');

%*******************************************************
% SSPA & TWTA using AWGN performance
 %Performance of QPSK-OFDM system using SSPA with
different IBO and K = 64.
%*******************************************************

clear;
tic

IBO_dB=0;           % backoff (dB)
target=1e-5;        % simulate to this BER
testing=0;          % testing=1 short run, testing=0
long run
alpha=4;            % oversampling factor

M=4;%[2 4 8 16];

for iM=1:length(M)

 %with PA
 for i=1:length(IBO_dB)

    [EbN0_dB BER]=BER_fun(target,testing,M(iM),
alpha,IBO_dB(i));
    data=[EbN0_dB BER]';
```

```
  % No PA
  %[EbN0_dB BER]=BER_fun(target,testing,M(iM),alpha,
'no thanks');
  %data=[EbN0_dB BER]';
end

end

semilogy(EbN0_dB,BER,'r')
grid on
axis([0 30 1e-5 1e0])
hold on
clc
tmp=toc/60
xlabel('Eb/No  dB')
ylabel('BER')

%*******************************************************
%BER_function
%*******************************************************

function [EbN0_dB BER]=BER_fun(target,testing,M,alpha,
IBO_dB)
% BER_fun.m
%
% input:

%target=1e-5; %- target BER
%testing=1;   % - 1=short run, 2=long run
%M=16;          % - modulation order
%alpha=4;       % - oversampling factor
%IBO_dB=3;       %- backoff (if char, no PA)
%
% output:
%   EbN0_dB - SNR (dB)
%   BER - bit error rate
%-----------------------------------------------------------------
% --------- Signal parameters ---------------------------

% bit mappings
tmp=(0:M-1)';
SymMap=exp(j*2*pi*tmp/M);  % data symbol mapping
if M==2
  BitMap=[0; 1];                % bit mapping
```

```
elseif M==4
  BitMap=[0 0; 0 1; 1 1; 1 0];
elseif M==8
  BitMap=[...
    0 0 0;
    0 0 1;
    0 1 1;
    0 1 0;
    1 1 0;
    1 1 1;
    1 0 1;
    1 0 0];
elseif M==16
  BitMap=[...
    0 0 0 0;
    0 0 0 1;
    0 0 1 1;
    0 0 1 0;
    0 1 1 0;
    0 1 1 1;
    0 1 0 1;
    0 1 0 0;
    1 1 0 0;
    1 1 0 1;
    1 1 1 1;
    1 1 1 0;
    1 0 1 0;
    1 0 1 1;
    1 0 0 1;
    1 0 0 0];
else error('M=2,4,8,16')
end
% /bit mappings

io=1;                   % index offset
N=64;                   % number of subcarriers
T=128e-6;               % symbol time
Fs=alpha*N/T;           % sampling rate
Ts=1/Fs;                % sampling period
Ns=T/Ts;                % samples per symbol
L=64;                   % vectorize.
Ntot=L*Ns;              % total samples per run

n=[0:Ns-1]';
W=zeros(Ns,N);
```

```
for k=0:N-1
  W(:,k+io)=exp(j*2*pi*k*n*Ts/T);
end


%
% --- PA parameters ----------------------------------
if ~ischar(IBO_dB)
  p=2;                      % PA linearity parameter
  g0=1;                     % gain (arbitrary)
  Asat=1;                   % input saturation level
(arbitrary)
  IBO=10^(IBO_dB/10);       % input backoff
  Pin=Asat^2/IBO;           % scale input power to Pin
  ph=pi/12;                 %pi/12 for TWTA
  ta=0.25;
end
%---------Simulation--------------------------------
%
if testing
  error_min=20;             % short run
  trans_min=1e4;            % min bits
else
  error_min=400;            % long run
  trans_min=1e5;            % min bits
end
trans_max=error_min/target; % max bits
%
BER=0;                      % initialize BER vector
EbN0_dB=0;                  % initialize SNR
iSNR=1;                     % SNR counter
go=1;                       % initialize loop
while go                    % run until max SNR
condition
  error_num=0; trans_num=0;  % initialize
  while trans_num<=trans_min | (error_num<=error_min &
trans_num<=trans_max)
    in=ceil(M*rand(N,L));      % random symbol index
    I=SymMap(in);              % data symbols
    s=W*I;                     % OFDM signal
    s=reshape(s,Ntot,1);       % to vector

    % PA
    if ~ischar(IBO_dB)
      Ps=1/(Ntot*Ts)*sum(abs(s).^2)*Ts; % input signal
power
```

```
       yin=s*sqrt(Pin/Ps);              % input signal with
backoff
       Ayin=abs(yin);                   % input envelope
       G=g0*Ayin./(1+(Ayin/Asat).^(2*p)).^(1/(2*p));
% AM/AM
       P=(ph*Ayin.^2)./(1+(ta*Ayin.^2));                      %
AM/PM
       yout=G.*exp(j*(angle(yin)+P)); % output signal
    else
       yout=s;
    end
    % /PA

    Eyout=sum(abs(yout).^2)*Ts; % signal energy
    Eb=Eyout/(L*N*log2(M));      % bit energy
    EbN0=10^(EbN0_dB(iSNR)/10); % SNR
    N0=Eb./EbN0;                 % noise spectral height
    dum=sqrt(1/2)*(randn(Ntot,1)+j*randn(Ntot,1));
    z=sqrt(N0*Fs)*dum;           % noise signal
    r=reshape(yout+z,Ns,L);      % received OFDM, to
matrix
    Ihat=W'*r/Ns;                % output of
matched-filter

    % detector
    errors=0;
    for iL=1:L                   % for each OFDM symbol
       % M copies of iL RX block
       dum1=Ihat(:,iL)*ones(1,M);
       % M columns, one per symbol element
       dum2=ones(N,1)*SymMap.';
       % distance to each point
       dum3=abs(dum1-dum2);
       % ix: vector w/ column index of min element in
each row of dum3
       % x: don't care
       [x,ix]=min(dum3');
       errors=errors+sum(sum(BitMap(ix,:)~=BitMap
(in(:,iL),:)));
    end
    % /detector

    error_num=error_num+errors; % cumulative bit errors
    trans_num=trans_num+L*N*log2(M); % cumulative
transmitted bits
```

```
  end % end this SNR

  BER(iSNR)=error_num/trans_num;
  % max SNR condition
  if BER(iSNR)<target|EbN0_dB(iSNR)>=45
    go=0;
  else
    iSNR=iSNR+1;
    EbN0_dB(iSNR)=EbN0_dB(iSNR-1)+1;
  end
end % end simulation
BER=BER';
EbN0_dB=EbN0_dB';

%********************************************************
%Class A power amplifier efficiency
%********************************************************

i=0;
for IBO=0:10;
    i=i+1;
ibo=10^(IBO/10);
eff(i)=0.5*(1./ibo)*100;
end
IBO=0:10;
plot(IBO,eff)
grid on
xlabel('Input Power Backoff, IBO dB')
ylabel('Class  A  PA  Efficiency, %')

%********************************************************
%power spectral density
%********************************************************

clear
nFFTSize = 64;
% for each symbol bits a1 to a52 are assigned to
subcarrier
% index [-26 to -1 1 to 26]
subcarrierIndex = [-26:-1 1:26];
nBit = 2500;
ip = rand(1,nBit) > 0.5; % generating 1's and 0's
nBitPerSymbol = 52;

nSymbol = ceil(nBit/nBitPerSymbol);
```

```
% BPSK modulation
% bit0 --> -1
% bit1 --> +1
ipMod = 2*ip - 1;
ipMod = [ipMod zeros(1,nBitPerSymbol*nSymbol-nBit)];
ipMod = reshape(ipMod,nSymbol,nBitPerSymbol);

st = []; % empty vector

for ii = 1:nSymbol

inputiFFT = zeros(1,nFFTSize);

% assigning bits a1 to a52 to subcarriers [-26 to -1,
1 to 26]
inputiFFT(subcarrierIndex+nFFTSize/2+1) = ipMod(ii,:);

%  shift subcarriers at indices [-26 to -1] to fft
input indices [38 to 63]
inputiFFT = fftshift(inputiFFT);

outputiFFT = ifft(inputiFFT,nFFTSize);

% adding cyclic prefix of 16 samples
outputiFFT_with_CP = [outputiFFT(49:64) outputiFFT];

st = [st outputiFFT_with_CP];

end

close all
fsMHz = 20;
[Pxx,W] = pwelch(st,[],[],4096,20);
plot([-2048:2047]*fsMHz/4096,10*log10(fftshift(Pxx)));
xlabel('frequency, MHz')
ylabel('power spectral density')
title('Transmit spectrum OFDM (based on 802.11a)');

%*******************************************************
%Capacity of the MIMO system for different number of
antennas when the channel is unknown at the
transmitter
%*******************************************************

% M=3; %-> number of antennas (M x M) system
```

```
% corr=0; %-> 1 if with correlation, 0 if uncorrelated
(for a 2x2 system only)
% value=0.5;% -> correlation coefficient value from 0
->1
% XPD=0 ;%-> 1 if antenna XPD is to be investigated, 0
if not (for a 2x2 system
% %only)
% alpha=0.5;% -> XPD value
% output='out';% -> defined by 'erg' and 'out' for
ergodic capacity or outage
% %capacity respectively

function z=capacity_plot_main(M,corr,value,XPD,alpha,
output)


%*******************************************************


M=3; %-> number of antennas (M x M) system
corr=0; %-> 1 if with correlation, 0 if uncorrelated
(for a 2x2 system only)
value=0.5;% -> correlation coefficient value from 0 ->1
XPD=0 ;%-> 1 if antenna XPD is to be investigated, 0
if not (for a 2x2 system
%only)
alpha=0.5;% -> XPD value
output='out';% -> defined by 'erg' and 'out' for
ergodic capacity or outage
%capacity respectively


%*******************************************************


%vary SNR through 20 dB
SNR=0:1:20;%SNR is signal-to-noise ratio in dBs
temp2=[];
for i=1:length(SNR)
temp1(i)=capacity_plot(SNR(i),M,corr,value,XPD,alpha,
output);% change this file to suit requirements e.g.
capacity_rician,

    temp2=[temp2 temp1(i)];
    temp1(i)=0;
end

plot(SNR,temp2,'r-^');
grid;
```

```
%plot routines follow. These will change depending
upon the type of plot.
%The following routines are based on the given example
above
xlabel('SNR');
ylabel('Capacity (Bits/sec)');
title('Outage Capacity Variation with SNR for Corr =
0.5 and XPD = 0.5');


%*****************************************************
%capacity_plot function
%*****************************************************

function output=capacity_plot(SNR,M,corr,value,XPD,alp
ha,output)

SNR=10^(0.1*SNR);

%10000 Monte-Carlo runs
for K=1:10000
   T=randn(M,M)+j*randn(M,M);
   T=0.707*T;
 if corr
     T=[1 value;value 1];
     T=chol(T);
 elseif XPD
     T=[1 alpha;alpha 1];
     T=chol(T);
 end
   I=eye(M);
   a=(I+(SNR/M)*T*T');
   a=det(a);
   y(K)=log2(a);
end
[n1 x1]=hist(y,40);
n1_N=n1/max(K);
a=cumsum(n1_N);
b=abs(x1);
if output == 'erg'
    output=interp1q(a,b',0.5);   %ergodic capacity
elseif output == 'out'
    output=interp1q(a,b',0.1);   %outage capacity
end
```

```
%**********************************************************
% In this program a MIMO channel with nt transmit
antenna and %nr receive
% antenna is analyzed.

%Capacity of the MIMO system for different number of
antennas %when the channel is known at the transmitter
%The power in parallel channel is distributed as
water-filling %algorithm
%**********************************************************

clear all
close all
clc

nt_V = [1 2 3 2 4];
nr_V = [1 2 2 3 4];

N0 = 1e-4;
B  = 1;
Iteration = 1e4;

SNR_V_db = [0:2:20];
SNR_V    = 10.^(SNR_V_db/10);

color = ['b';'r';'g';'k';'c'];
notation = ['-o';'->';'<-';'-^';'-s'];

for(k = 1 : 5)
    nt = nt_V(k);
    nr = nr_V(k);
    for(i = 1 : length(SNR_V))
        Pt = N0 * SNR_V(i);
        for(j = 1 : Iteration)
            H = random('rayleigh',1,nr,nt);
            [S V D] = svd(H);
            landas(:,j)  = diag(V);
            [Capacity(i,j) PowerAllo] =
WaterFilling_alg(Pt,landas(:,j),B,N0);
        end
    end

    f1 = figure(1);
    hold on
```

```
    plot(SNR_V_db,mean(Capacity'),notation(k,:),'color',
color(k,:))

    clear landas
end

legend_str = [];
for( i = 1 : length(nt_V))
    legend_str =[ legend_str ;...
        {['nt = ',num2str(nt_V(i)),' , nr =
',num2str(nr_V(i))]}];
end
legend(legend_str)
grid on
set(f1,'color',[1 1 1])
xlabel('SNR in dB')
ylabel('Capacity bits/s/Hz')

%*********************************************************
% WaterFilling in Optimizing the Capacity
%WaterFilling_alg function
%*********************************************************

function [Capacity PowerAllo]=WaterFilling_alg
(PtotA,ChA,B,N0);

% Initialization
%===============
ChA = ChA + eps;
NA = length(ChA);       % the number of subchannels
allocated to

H = ChA.^2/(B*N0);   % the parameter relate to SNR in
subchannels
% assign the power to subchannel
PowerAllo = (PtotA + sum(1./H))/NA - 1./H;
while(length(find(PowerAllo < 0 ))>0)
    IndexN = find(PowerAllo <= 0 );
    IndexP = find(PowerAllo > 0);
    MP = length(IndexP);
    PowerAllo(IndexN) = 0;
    ChAT = ChA(IndexP);
    HT = ChAT.^2/(B*N0);
    PowerAlloT = (PtotA + sum(1./HT))/MP - 1./HT;
```

```
        PowerAllo(IndexP) = PowerAlloT;
end
PowerAllo = PowerAllo.';
Capacity  = sum(log2(1+ PowerAllo.' .* H));


%*******************************************************
%%% Alamouti Space Time Code

% It is assumed that system contains 2 transmit
antennas and one receive
% antenna. At the transmitter the data of a two
consecutive slots will be
% considered. At the odd time slots, the first antenna
transmit
% symbol 1 (s1) and the second ones will transmit
symbol 2 (s2)
% simultaneously. At the even time slots the -s2* and
s1* will be transmitted
% from the first and second antenna respectively.
%*******************************************************

% Initialization
    clear
    clc

% Setting parameters
    numOfBlk = 1e6;      % number of blocks of data to
be transmitted
    qamOrder = 16;       % the QAM modulation order
4,16,64
    SNRdB    = 0:2:30;

    linColor = 'b'; % graph color
    linSym   = 'o'; % graph Symbol

% Memory allocation
errRate = zeros(size(SNRdB));

%%% AlamoutiSpace Time Code
for i = 1 : length(SNRdB)

% Main Program
    % generating the data
    txData = randint(numOfBlk*2,1,qamOrder);
```

```
    % splitting the data into two vectors (first
transmition, second
    % transmition in time);
    temp = reshape(txData,numOfBlk,2);

    % QAM Modulation of transmit data
    temp = qammod(temp,qamOrder);

    % 2 transmit antenna and 1 receive antenna channel
gain, the channel
    % variance is set to unity and a Rayleigh flat
fading channel in each
    % path is assumed
    H   = 1/sqrt(2) * (randn(numOfBlk,2) +
sqrt(-1)*randn(numOfBlk,2));

    % transmitted data through channel
    % in each transmit antenna half of the power will
be sent
    % 1/sqrt(2) is to represent the half of the power
on each antenna
    txMod(:,1) =  H(:,1).* 1/sqrt(2).*temp(:,1)      +
H(:,2).* 1/sqrt(2).*temp(:,2)       ;
    txMod(:,2) = -H(:,1).*(1/sqrt(2).*temp(:,2)').' +
H(:,2).*(1/sqrt(2).*temp(:,1)').' ;

    % adding noise
    txMod = awgn(txMod,SNRdB(i),'measured');

    % receiving the data
    % sqrt(2) is used for normalization
    temp(:,1) = sqrt(2)*(H(:,1)'.' .* txMod(:,1) +
H(:,2) .* txMod(:,2)').')./(abs(H(:,1)).^2 +
abs(H(:,2)).^2);
    temp(:,2) = sqrt(2)*(H(:,2)'.' .* txMod(:,1) -
H(:,1) .* txMod(:,2)').')./(abs(H(:,1)).^2 +
abs(H(:,2)).^2);

    rxData(:,1) = qamdemod(temp(:,1),qamOrder);
    rxData(:,2) = qamdemod(temp(:,2),qamOrder);

    [numErr errRate(i)] = biterr(rxData,reshape(txData,
numOfBlk,2));
end
```

```
    f1 = figure(1);
    semilogy(SNRdB,errRate,[linColor,'-',linSym]);
    xlabel('SNR in dB');
    ylabel('Symbol Error Rate');

%%%%%%%%%%%%%%%% No space time coding %%%%%%%%%%%%%%%%

for i = 1 : length(SNRdB)
    % txData is set in the Alamouti Code
    temp = qammod(txData,qamOrder);

    % Channel Definition
    H  = 1/sqrt(2) * (randn(numOfBlk*2,1) +
sqrt(-1)*randn(numOfBlk*2,1));

    % passing through channel
    txMod = H.*temp;

    % adding noise
    txMod = awgn(txMod,SNRdB(i),'measured');

    % decoding
    temp = txMod./H;
    rxData = qamdemod(temp,qamOrder);

    [numErr errRate(i)] = biterr(rxData,txData);
end

    figure(1);
    hold on
    semilogy(SNRdB,errRate,[linColor,':',linSym]);
    xlabel('SNR in dB');
    ylabel('Bit Error Rate');
    legend('With Alamouti Code 2x1 chann','No STC 1x1
chann')
    grid on

%****************************************************
% This program simulates the 2xnr MIMO-STBC system with %
%%%%%%%%%%%%%%%%%%  Alamouti code    %%%%%%%%%%%%%%%%%%%
%****************************************************

clc;
clear all;
```

```
%%%%%%%%%%%%%%%%%%%%%%%%%%%%%%%%%%%%%%%%%%%%%%%%%%%
for w=1:4;
    if w==4
        Rx=3;  % nr=1,2,4,6
    else
        Rx=2.^(w-1);
    end
%%%%%%%%%%%%%%%%%%%%%%%%%%%%%%%%%%%%%%%%%%%%%%%%%%%

 M=8;   %M=input('The modulation array scheme M-PSK
M = ');  %M=4 for 4PSk and 8 for 8PSK
 SNR_MAX=20;   %input('Enter the max SNR of the System
SNR_MAX = ');
 N=100000;     %input('Enter the Number of transmitted
symbol N = ');
               %Nr=input('Enter the number of receiving
antenna Nr = ');
randn('state',0);
%rand('state',0);


%$$$$$$$$$$$$$$$$$$$$$$$$ transmitter %$$$$$$$$$$$$$$$$$$$$$$

z=0;
for k=0:2:SNR_MAX
    A=floor(M*rand(2,N));            % transmitted
alphabet
    Str=exp(j*2*pi/M*A)/sqrt(2);   % Transmitted symbol
    for i=1:N
        S=[Str(1,i);Str(2,i)];   % making space-time
coding matrix
        snr=10.^(k/10);
        sig=(0.5)/snr;
        Ns=sqrt(sig).*(randn(2.*Rx,1)+j*(randn(2.
*Rx,1))); % noise matrix
        H=[];
%$$$$$$$$$$$$$$$$$$$$$$$$$$$$$$$$$   channel   $$$$$$$$$$$$$$$
        for p=1:Rx
            h=(randn(1,2)+j*(randn(1,2)))/sqrt(2);
            H=[H;h(1) h(2);h(2)' -h(1)']; % channel
matrix
        end
        p=0;
        r=H*S+Ns; % $$$$$$$$$$$$$$$ received signal
$$$$$$$$$$$$$$$$$$$$$$$$
```

```
        S_=H'*r;
        ang=angle(S_); % received angles
        B=mod(round(ang/(2*pi/M)),M);    % received
alphabet
        Ses(:,i)=[B(1);B(2)];            % received
symbol
    end
 %$$$$$$$$$$$$$$$$$$$$$$$$$$$$ BER calculation  $$$$$$$
$$$$$$$$$$$$$$$$$$$$$$$

  BER=0;
  ERROR=0;
    for p=1:2
        for i=1:N
          if A(p,i)~=Ses(p,i)
              ERROR=ERROR+1;
          end
        end
  end
  z=z+1;
  ber(z)=ERROR/(2*N);
end
snr=0:2:SNR_MAX;
ber;
Ber(:,w)=ber';
end
semilogy(snr,Ber(:,1),'b',snr,Ber(:,2),'g',snr,Ber(:,3),
'r',snr,Ber(:,4),'m');
grid on
hold on

%*****************************************************
                %%%%MIMO OFDM Simulation
%*****************************************************

clear all;
clc;
fprintf('Start! Please wait to inspect the results
...\n\n');

% Initial Processing:
% Define the slot structure
Nc = 512;
Ng = 32;
Ns = Nc + Ng;
```

```
Nu = Nc;
Num_Block = 1;
Fs = 20e6;
T = 1/Fs;
Tg = T * Ng;
Tu = T * Nc;
Ts = T * Ns;
DeltaF = 1/Tu;
B = DeltaF*Nu;
ModScheme = '16QAM';
M = 16;
Num_TxAnt = 2;
Num_RxAnt = 2;

Num_Bit_Frame = Num_Block * Nu * log2( M ) *
Num_TxAnt;
Num_Sym_Frame = Num_Bit_Frame / log2(M);

Gen_Poly = [13 15];
Len_Constr = 4;
Len_Mem = 3;
k = 1;
n = 3;
Rate = k / n;
Trellis = poly2trellis( Len_Constr,Gen_Poly,Gen_
Poly(1) );
Alg = 1;
Num_Iter_Decode = 8;
Num_InforBit = 2048;
Num_CodeBit = Num_InforBit / Rate + Len_Mem * 4;
Puncture_Pattern = [1 1; 1 0; 0 1];
[1; 1; 1];
[1 1; 1 0; 0 1];
[1 1 1 1; 1 0 0 0; 0 0 1 0];
[1 1 1 1 1 1; 1 0 0 0 0 0; 0 0 0 1 0 0];
[1 1 1 1 1 1 1 1 1 1; 1 0 0 0 1 0 0 1 0 0; 0 0 1 0 0 1
0 0 0 1];
Len_Pattern = prod( size( Puncture_Pattern ) );
Num_Reserved = sum( sum( Puncture_Pattern,1 ),2 );
Num_Punctured =  Len_Pattern - Num_Reserved;
Rate = size( Puncture_Pattern,2 ) / Num_Reserved;

Puncture_Pattern = reshape( Puncture_Pattern,1,Len_
Pattern ) ;
```

```
if ( Num_Bit_Frame - fix( Num_Bit_Frame / Num_Reserved )
* Num_Reserved ) == 0;
    Num_CodeBit_Frame = fix( Num_Bit_Frame / Num_
Reserved ) * Len_Pattern;
else
    for i = 1 : Len_Pattern
        if sum( Puncture_Pattern( 1:i ) ) == ( Num_
Bit_Frame - fix( Num_Bit_Frame / Num_Reserved )
* Num_Reserved )
            Num_CodeBit_Frame = fix( Num_Bit_Frame /
Num_Reserved ) * Len_Pattern + i;
        end
    end
end
while mod( Num_CodeBit_Frame,Num_CodeBit ) ~= 0
    Num_InforBit = Num_InforBit - 1;    Num_CodeBit =
Num_InforBit * n / k + Len_Mem * 4;
end
Num_CodeBlock = Num_CodeBit_Frame / Num_CodeBit;
Num_InforBit_Frame = Num_InforBit * Num_CodeBlock;
Rate_Source = Num_InforBit_Frame / Num_Bit_Frame;
[Temp, Inner_Interlver] = sort( rand( 1,Num_
InforBit ) );
Inner_Interlver = Inner_Interlver -1;
[Temp, Outer_Interlver] = sort( rand( 1,Num_Bit_Frame ) );

%///////// Define the channel profile

 % Path_Gain = [ 1 ];  Path_Delay = [0];
    %ChannelProfile = 'AWGN';

% Path_Gain = [0.9977    0.0680];    Path_Delay = [0 2];
   %ChannelProfile = 'ITU Pedestrian A';

  % Path_Gain = [0.6369 0.5742 0.3623 0.2536 0.2595
0.0407];    Path_Delay = [0 1 4 6 11 18];
    %ChannelProfile = 'ITU Pedestrian B';

Path_Gain = [0.6964 0.6207 0.2471 0.2202 0.1238
0.0696];
Path_Delay = [0 1 2 3 4 5] + 1;
ChannelProfile = 'ITU Vehicular A';
%Path_Gain = [0.4544 0.4050 0.3610 0.3217 0.2867
0.2555 0.2277 0.2030 0.1809 0.1612 0.1437 0.1281...
```

```
%                0.1141 0.1017 0.0907 0.0808 0.0720
0.0642 0.0572 0.0510 0.0454 0.0405 0.0361 0.0322];
%Path_Delay = [0 7 14 22 29 37 45 52 59 67 75 82 90 97
104 112 119 127 135 142 150 157 164 172] + 1;
   %ChannelProfile = 'Exponential Decay Model';

Num_Path = length( Path_Gain );
Max_Delay = max( Path_Delay );
Fc = 3e9;
V = 3;
Fd = V * Fc / 3e8 * 1000 / 3600;
Phase = 2 * pi * rand( 1,Num_Path*Num_RxAnt*Num_TxAnt );

% Save simulation parameters
 FileName = 'Sim_MIMO_OFDM.dat';
% % % FileName = 'Sim_MIMO_OFDM.dat';
% Fid = fopen(FileName,'a+'); fprintf(Fid,'\n\n');
fprintf(Fid,['%% Created by ZZG from <' mfilename '.m>
at ' datestr(now),'\n']);
% fprintf(Fid,'%% Num_Path = %d  vehicle speed = %d
carrier frequency = %e  Doppler frequency spread = %f
normalized Doppler shift =
%f\n',Num_Path,V,Fc,Fd,Fd*Ts);
% fprintf(Fid,'%% system bandwidth = %e  number of
subcarriers = %d  subcarrier spacing =
%e\n',B,Nc,DeltaF);
% fprintf(Fid,'%% sampling duration = %e  symbol
duration = %e  guard duration = %e \n',T,Ts,Tg);
% fprintf(Fid,'%% (%d, %d, %d)  Generator = %s  Num_
InforBit = %d  Num_CodeBlock = %d  Num_InforBit_Frame
= %d  Rate = %f  \n',n,k,Len_Constr,num2str( Gen_Poly
),Num_InforBit,Num_CodeBlock,Num_InforBit_Frame,Rate);
% fprintf(Fid,'%% Num_Block = %d  ModScheme = %s  Num_
TxAnt = %d  Num_RxAnt =
%d\n',Num_Block,ModScheme,Num_TxAnt,Num_RxAnt);
% fprintf(Fid,'%% channel profile =
%s\n',ChannelProfile );
% fprintf(Fid,'%% SNR  BER  FER \n\n'); fclose(Fid);

% [(0 : 1 : 3) (4 : 0.5 : 6)]
% Main loop
%SNR = [( 1:1:12 ) ];
SNR = [( 1:1:9 ) ];
MinSNR = min(SNR);
MaxSNR = max(SNR);
```

```
BER = [];
FER = [];
Num_Iter = 6;
Num_Frame = 10;
for Index = 1 : length( SNR )
    %    profile on -detail builtin
    StartPoint = 0;
    snr = SNR( Index )
    EbN0 = 10^( snr / 10 );
    Es = 1;
    N0  = Es * Num_RxAnt / ( EbN0 * Rate * Nu/Ns *
log2(M) * Num_TxAnt );
    Var = N0;
    ErrNum_Bit = zeros( 1,Num_Iter );
    ErrNum_Frame = zeros( 1,Num_Iter );
    ErrRate_Bit = zeros( 1,Num_Iter );
    ErrRate_Frame = zeros( 1,Num_Iter );

    for Frame = 1 : Num_Frame
        tic;
        %///////////// Transmitter //////////////////

        Data_In = randint( 1,Num_InforBit_Frame );
        for i = 1 : Num_CodeBlock
            % Data_EnCode( (i-1)*Num_CodeBit+(1:Num_
CodeBit) ) = Enc_Conv( Data_In( (i-1)*Num_InforBit+
(1:Num_InforBit) ),Trellis,InitState,Terminated );
            Data_EnCode( (i-1)*Num_CodeBit+(1:Num_
CodeBit) ) = Enc_Turbo_3gpp( Data_In( (i-1)*Num_InforBit+
(1:Num_InforBit) ),Gen_Poly,Len_Constr,Inner_Interlver );
        end
        Data_EnCode = Puncture( Data_EnCode,Puncture_
Pattern );
        Sym_In = reshape( Mapping( Data_EnCode( Outer_
Interlver ),ModScheme ),Num_TxAnt,Nu*Num_Block ) /
sqrt( Num_TxAnt );
        for TxAnt = 1 : Num_TxAnt
            Temp = reshape( Sym_In( TxAnt,: ),Nc,Num_
Block );
            Temp = ifft( Temp,Nc,1 ) * sqrt( Nc );
            TransSig( TxAnt,: ) = reshape(
[Temp( Nc-Ng+1:Nc,: );Temp],1,Ns*Num_Block );
        end

        % /////////////Channel//////////////////////
```

```
        ChannelCoeff = MultiPathChannel( repmat( Path_
Gain,1,Num_RxAnt*Num_TxAnt ),Fd,Ts,Num_
Block,StartPoint,Phase );  StartPoint = StartPoint +
Num_Block;
        % ChannelCoeff = diag( repmat( Path_
Gain,1,Num_RxAnt*Num_TxAnt ) ) * ( randn( Num_
Path*Num_RxAnt*Num_TxAnt,Num_Block ) + sqrt( -1 ) *
randn( Num_Path*Num_RxAnt*Num_TxAnt,Num_Block ) ) /
sqrt( 2 );
        ChannelOut = zeros( Num_RxAnt,Ns*Num_
Block+Max_Delay-1 );
        for RxAnt = 1 : Num_RxAnt
            for TxAnt = 1 : Num_TxAnt
                h( Path_Delay,1:Num_Block ) =
ChannelCoeff( (RxAnt-1)*Num_Path*Num_TxAnt + (TxAnt-
1)*Num_Path + (1:Num_Path) ,:);
                H( RxAnt,TxAnt,: ) = reshape( fft(
h,Nc,1 ),1,Nc*Num_Block );
                for i = 1 : Num_Block
                    Temp = ChannelOut( RxAnt, (i-1)*Ns
+ (1:Ns+Max_Delay-1) );
                    ChannelOut( RxAnt,(i-1)*Ns +
(1:Ns+Max_Delay-1) ) = Temp + conv( h(:,i),TransSig(
TxAnt,(i-1)*Ns + (1:Ns) ) );
                end
            end
        end
        RecSig = ChannelOut + sqrt( Var ) * ( randn(
size( ChannelOut ) ) + sqrt( -1 ) * randn( size(
ChannelOut ) ) ) / sqrt( 2 );
        clear ChannelCoeff h ChannelOut;

        %///////////// Receiver/////////////

        RecSig = RecSig( :,1:Ns*Num_Block );
        for RxAnt = 1 : Num_RxAnt
            Temp = reshape( RecSig( RxAnt,: ),Ns,Num_
Block );
            RecSig_Fre( RxAnt,: ) = reshape( fft(
Temp( Ng+1:Ns,: ) ) / sqrt( Nc ),1,Nc*Num_Block );
        end
        Y = RecSig_Fre;
        HH = H / sqrt( Num_TxAnt );
        clear RecSig RecSig_Fre H;
```

```
        Lu_Pri = zeros( 1,Num_InforBit_Frame );
        Lc_Pri = zeros( 1,Num_Bit_Frame );
        for Iter = 1 : Num_Iter
            Lc_Extr = MMSE_Equ(
Y,HH,Lc_Pri,Num_RxAnt,Num_TxAnt,ModScheme,Var,1);
            DeInterlv( Outer_Interlver ) = Lc_Extr;
            Lc_Pri = DePuncture( DeInterlv,Num_
CodeBit_Frame,Puncture_Pattern );
            for i = 1 : Num_CodeBlock
                % [Temp, Lc_Extr( (i-1)*Num_
CodeBit+(1:Num_CodeBit) )] = ......
                % LOG_MAP( zeros( 1,Num_InforBit +
Len_Mem ),Lc_Pri( (i-1)*Num_CodeBit+(1:Num_CodeBit) ),
Trellis,1 );
                % Data_Out( (i-1)*Num_
InforBit+(1:Num_InforBit) ) = ( sign( Temp( 1:Num_
InforBit ) ) + 1 ) / 2;
                [Data_Out( (i-1)*Num_
InforBit+(1:Num_InforBit) ),Lu_Extr( (i-1)*Num_
InforBit+(1:Num_InforBit) ),Lc_Extr(
(i-1)*Num_CodeBit+(1:Num_CodeBit) )] = ......
                    Dec_Turbo_3gpp( Lu_Pri(
(i-1)*Num_InforBit+(1:Num_InforBit) ),Lc_Pri(
(i-1)*Num_CodeBit+(1:Num_CodeBit) ),Trellis,Inner_
Interlver,Alg,Num_Iter_Decode );
            end
            Lc_Extr = Puncture( Lc_Extr,Puncture_
Pattern );
            Lc_Pri = Lc_Extr( Outer_Interlver );
            Error = sum( sum( sign( abs( Data_Out
- Data_In ) ) ) );
            ErrNum_Bit( 1,Iter ) = ErrNum_Bit( 1,Iter
) +  Error;
            if ( Error ~= 0 )
                ErrNum_Frame( 1,Iter ) = ErrNum_Frame(
1,Iter ) +  1;
            end
            ErrRate_Bit( 1,Iter ) = ErrNum_Bit( 1,Iter )
/ Frame / Num_InforBit_Frame;
            ErrRate_Frame( 1,Iter ) = ErrNum_
Frame( 1,Iter ) / Frame;
        end
        Frame
        ErrRate_Bit
```

```
        ErrRate_Frame
        toc;
    end
%    profile report
    BER = [BER; ErrRate_Bit];
    FER = [FER; ErrRate_Frame];
    Fid = fopen(FileName,'a+');
    fprintf(Fid,'%2.1f        %s
%s\n',snr,num2str( ErrRate_Bit,'%1.10f ' ),num2str(
ErrRate_Frame,'%1.10f ' ) );
    fclose(Fid);
end

figure(1)
semilogy(SNR,BER,SNR,FER);
axis([MinSNR,MaxSNR,10^-6,10^0]);
grid on;
xlabel('Eb/N0 (dB)');  ylabel('BER')
%
figure(2)
Throughput = (1 - FER) * Num_InforBit_Frame/Num_Block/
Ts;
plot(SNR,Throughput/1e6);
axis([MinSNR,MaxSNR,0,110]);
grid on;
xlabel('Eb/N0 (dB)');  ylabel('Throughput (Mbps)')

%*******************************************************
%
% Simulation program to simulate OFDM transmission system
%
%*******************************************************
%**************** preparation part *****************
%*******************************************************

l=0;
for ebn0=0:2:20
l=l+1;
para=128;    % Number of parallel channel to transmit
(points)
fftlen=128; % FFT length
noc=128;    % Number of carrier
nd=6;       % Number of information OFDM symbol for
one loop
ml=2;       % Modulation level : QPSK
```

```
sr=250000;  % Symbol rate
br=sr.*ml;  % Bit rate per carrier
gilen=32;   % Length of guard interval (points)
%ebn0=3;    % Eb/N0


%******************************************************
%****************** main loop part ******************
%******************************************************


nloop=100;  % Number of simulation loops

noe = 0;    % Number of error data
nod = 0;    % Number of transmitted data
eop=0;      % Number of error packet
nop=0;      % Number of transmitted packet

for iii=1:nloop

%******************************************************
%**************** transmitter *********************
%******************************************************


%*************// (1) Data generation //**************

seldata=rand(1,para*nd*ml)>0.5;  %  rand : built in
function

%********// (2)Serial to parallel conversion //*******

paradata=reshape(seldata,para,nd*ml); %  reshape :
built in function

%*************// (3)QPSK modulation //***************

[ich,qch]=qpskmod(paradata,para,nd,ml);
kmod=1/sqrt(2); %  sqrt : built in function
ich1=ich.*kmod;
qch1=qch.*kmod;

%*****************// (4)IFFT //********************

x=ich1+qch1.*i;
y=ifft(x);       %  ifft : built in function
ich2=real(y);    %  real : built in function
qch2=imag(y);    %  imag : built in function
```

```
%*********//  (5)Gurad interval insertion  //*********

[ich3,qch3]= giins(ich2,qch2,fftlen,gilen,nd);
fftlen2=fftlen+gilen;

%***************  Attenuation Calculation  ***********

spow=sum(ich3.^2+qch3.^2)/nd./para;  %  sum : built in
function
attn=0.5*spow*sr/br*10.^(-ebn0/10);
attn=sqrt(attn);

%*******************************************************
%********************  Receiver  ********************
%*******************************************************

%**************  ////  (1)AWGN addition ////***********

[ich4,qch4]=comb(ich3,qch3,attn);

%*********  ////  (2)Guard interval removal ////********

[ich5,qch5]= girem(ich4,qch4,fftlen2,gilen,nd);

%***************////   (3)FFT  ////********************

rx=ich5+qch5.*i;
ry=fft(rx);     % fft : built in function
ich6=real(ry);  % real : built in function
qch6=imag(ry);  % imag : built in function

%************* ////(4)demoduration ////***************

ich7=ich6./kmod;
qch7=qch6./kmod;
[demodata]=qpskdemod(ich7,qch7,para,nd,ml);

%*****////  (5)Parallel to serial conversion  ////*****

demodata1=reshape(demodata,1,para*nd*ml);

%*******************************************************
%***************  Bit Error Rate (BER)  ***************
%*******************************************************
```

```
% instantaneous number of error and data

noe2=sum(abs(demodata1-seldata));  %  sum : built in
function
nod2=length(seldata);  %  length : built in function

% cumulative the number of error and data in noe and
nod

noe=noe+noe2;
nod=nod+nod2;

% calculating PER

if noe2~=0
   eop=eop+1;
else
   eop=eop;
end
   eop;
   nop=nop+1;

end
ber(l) = noe/nod
per=eop/nop;
end


%************************************************************
%***************** Output result ********************
%************************************************************

ebn=0:2:20;
semilogy(ebn,ber,'r')
axis([0 60 1e-4 1e0])
grid on
xlabel('Eb/No (dB)')
ylabel('BER')


%***************** end of file ********************

% Function to perform QPSK modulation

function [iout,qout]=qpskmod(paradata,para,nd,ml)
```

```
%****************** variables ***********************
% paradata : input data (para-by-nd matrix)
% iout :output Ich data
% qout :output Qch data
% para    : Number of parallel channels
% nd : Number of data
% ml : Number of modulation levels
% (QPSK ->2  16QAM -> 4)
% *****************************************************


m2=ml./2;

paradata2=paradata.*2-1;
count2=0;

for jj=1:nd

    isi = zeros(para,1);
    isq = zeros(para,1);

    for ii = 1 : m2
        isi = isi + 2.^( m2 - ii ) .*
paradata2((1:para),ii+count2);
        isq = isq + 2.^( m2 - ii ) .* paradata2((1:para),
m2+ii+count2);
    end

    iout((1:para),jj)=isi;
    qout((1:para),jj)=isq;

    count2=count2+ml;

end

%***************** end of file ***********************

% Function giins.m
%
% Function to insert guard interval into transmission
signal
%

function [iout,qout]= giins(idata,qdata,fftlen,
gilen,nd);
```

```
%***************** variables *********************
% idata    : Input Ich data
% qdata    : Input Qch data
% iout     : Output Ich data
% qout     : Output Qch data
% fftlen   : Length of FFT (points)
% gilen    : Length of guard interval (points)
% ********************************************************

idata1=reshape(idata,fftlen,nd);
qdata1=reshape(qdata,fftlen,nd);
idata2=[idata1(fftlen-gilen+1:fftlen,:); idata1];
qdata2=[qdata1(fftlen-gilen+1:fftlen,:); qdata1];

iout=reshape(idata2,1,(fftlen+gilen)*nd);
qout=reshape(qdata2,1,(fftlen+gilen)*nd);

%**************** end of file ********************

% Program comb.m
%
% Generate additive white Gaussian noise

%

function [iout,qout] = comb (idata,qdata,attn)

%***************** variables *********************
% idata : input Ich data
% qdata : input Qch data
% iout    output Ich data
% qout    output Qch data
% attn : attenuation level caused by Eb/No or C/N
%*********************************************************

iout = randn(1,length(idata)).*attn;
qout = randn(1,length(qdata)).*attn;

iout = iout+idata(1:length(idata));
qout = qout+qdata(1:length(qdata));

% *******************end of file********************

% Function to remove guard interval from received signal
```

```
function [iout,qout]= girem(idata,qdata,fftlen2,
gilen,nd);

%***************** variables **********************
% idata       : Input Ich data
% qdata       : Input Qch data
% iout        : Output Ich data
% qout        : Output Qch data
% fftlen2     : Length of FFT (points)
% gilen       : Length of guard interval (points)
% nd          : Number of OFDM symbols
% ****************************************************
idata2=reshape(idata,fftlen2,nd);
qdata2=reshape(qdata,fftlen2,nd);

iout=idata2(gilen+1:fftlen2,:);
qout=qdata2(gilen+1:fftlen2,:);

%***************** end of file **********************
%
% Function to perform QPSK demodulation
%


%

function [demodata]=qpskdemod(idata,qdata,para,nd,ml)

%***************** variables **********************
% idata :input Ich data
% qdata :input Qch data
% demodata: demodulated data (para-by-nd matrix)
% para    : Number of parallel channels
% nd : Number of data
% ml : Number of modulation levels
% (QPSK ->2  16QAM -> 4)
% ****************************************************

demodata=zeros(para,ml*nd);
demodata((1:para),(1:ml:ml*nd-1))=idata((1:para),
(1:nd))>=0;
demodata((1:para),(2:ml:ml*nd))=qdata((1:para),
(1:nd))>=0;

%***************** end of file **********************
```

```
% **************************************************

% Simulation program to realize OFDM transmission system
% (under one path fading)
% **************************************************

%**************** preparation part *****************
j=0;
for ebn0=0:2:42
    j=j+1;
para=128;    % Number of parallel channel to transmit
(points)
fftlen=128; % FFT length
noc=128;     % Number of carrier
nd=6;        % Number of information OFDM symbol for
one loop
ml=2;        % Modulation level : QPSK
sr=250000;   % Symbol rate
br=sr.*ml;   % Bit rate per carrier
gilen=32;    % Length of guard interval (points)
%ebn0=10;     % Eb/N0

%************** Fading initialization ***************

% Time resolution

tstp=1/sr/(fftlen+gilen);

% Arrival time for each multipath normalized by tstp
% If you would like to simulate under one path fading
model, you have only to set
% direct wave.

itau = [0];

% Mean power for each multipath normalized by direct
wave.
% If you would like to simulate under one path fading
model, you have only to set
% direct wave.
dlvl = [0];

% Number of waves to generate fading for each
multipath.
```

```
% In normal case, more than six waves are needed to
generate Rayleigh fading
n0=[6];

% Initial Phase of delayed wave
% In this simulation four-path Rayleigh fading are
considered.
th1=[0.0];

% Number of fading counter to skip
itnd0=nd*(fftlen+gilen)*10;

% Initial value of fading counter
% In this simulation one-path Rayleigh fading are
considered.
% Therefore one fading counter are needed.

itnd1=[1000];

% Number of direct wave + Number of delayed wave
% In this simulation one-path Rayleigh fading are
considered
now1=1;

% Maximum Doppler frequency [Hz]
% You can insert your favorite value
fd=340;

% You can decide two modes to simulate fading by
changing the variable flat
% flat     : flat fading or not
% (1->flat (only amplitude is fluctuated),0->normal
(phase and amplitude are fluctuated)
flat =1;

%****************** main loop part ******************

nloop=500;  % Number of simulation loops

noe = 0;    % Number of error data
nod = 0;    % Number of transmitted data
eop=0;      % Number of error packet
nop=0;      % Number of transmitted packet

for iii=1:nloop
```

```
%***************** transmitter *********************

%***************** Data generation ******************

seldata=rand(1,para*nd*ml)>0.5;  %  rand : built in
function

%*********** Serial to parallel conversion **********

paradata=reshape(seldata,para,nd*ml); %  reshape :
built in function

%***************** QPSK modulation *******************

[ich,qch]=qpskmod(paradata,para,nd,ml);
kmod=1/sqrt(2); %  sqrt : built in function
ich1=ich.*kmod;
qch1=qch.*kmod;

%****************** IFFT **********************

x=ich1+qch1.*i;
y=ifft(x);       %  ifft : built in function
ich2=real(y);    %  real : built in function
qch2=imag(y);    %  imag : built in function

%********* Gurad interval insertion **********

[ich3,qch3]= giins(ich2,qch2,fftlen,gilen,nd);
fftlen2=fftlen+gilen;

%********* Attenuation Calculation *********

spow=sum(ich3.^2+qch3.^2)/nd./para;  %  sum : built in
function
attn=0.5*spow*sr/br*10.^(-ebn0/10);
attn=sqrt(attn);

%***************** Fading channel *******************

% Generated data are fed into a fading simulator
[ifade,qfade]=sefade(ich3,qch3,itau,dlvl,th1,n0,itnd1,
now1,length(ich3),tstp,fd,flat);

% Update fading counter
```

```
itnd1 = itnd1+ itnd0;

%******************** Receiver ********************
%**************** AWGN addition *********

[ich4,qch4]=comb(ifade,qfade,attn);

%***************** Guard interval removal *********

[ich5,qch5]= girem(ich4,qch4,fftlen2,gilen,nd);

%***************** FFT *****************

rx=ich5+qch5.*i;
ry=fft(rx);      % fft : built in function
ich6=real(ry);   % real : built in function
qch6=imag(ry);   % imag : built in function

%**************** demoduration ******************

ich7=ich6./kmod;
qch7=qch6./kmod;
[demodata]=qpskdemod(ich7,qch7,para,nd,ml);

%********* Parallel to serial conversion **********

demodata1=reshape(demodata,1,para*nd*ml);

%*************** Bit Error Rate (BER) ***************

% instantaneous number of error and data
noe2=sum(abs(demodata1-seldata));  %  sum : built in
function
nod2=length(seldata);  %  length : built in function

% cumulative the number of error and data in noe and nod
noe=noe+noe2;
nod=nod+nod2;

% calculating PER
if noe2~=0
   eop=eop+1;
else
   eop=eop;
end
```

```
   eop;
   nop=nop+1;

%fprintf('%d\t%e\t%d\n',iii,noe2/nod2,eop);   %
fprintf : built in function


end

ber(j) = noe/nod
per=eop/nop;


end
%**************** Output result ********************

ebn=0:2:42;
semilogy(ebn,ber,'r')
axis([0 60 1e-4 1e0])
grid on
xlabel('Eb/No (dB)')
ylabel('BER')


%**************** end of file **********************

% This function generates frequency selecting fading
%

function[iout,qout,ramp,rcos,rsin]=sefade(idata,qdata,
itau,dlvl,th,n0,itn,n1,nsamp,tstp,fd,flat)


%**************** variables ***********************
% idata    input Ich data
% qdata    input Qch data
% iout     output Ich data
% qout     output Qch data
% ramp   : Amplitude contaminated by fading
% rcos   : Cosine value contaminated by fading
% rsin   : Cosine value contaminated by fading
% itau   : Delay time for each multipath fading
% dlvl   : Attenuation level for each multipath fading
% th     : Initialized phase for each multipath fading
% n0     : Number of waves in order to generate each
multipath fading
% itn    : Fading counter for each multipath fading
% n1     : Number of summation for direct and delayed
waves
```

```
% nsamp   : Total number od symbols
% tstp    : Minimum time resolution
% fd    : Maximum Doppler frequency
% flat      flat fading or not
% (1->flat (only amplitude is fluctuated),0->normal
(phase and amplitude are fluctuated)
%*********************************************************

iout = zeros(1,nsamp);
qout = zeros(1,nsamp);

total_attn = sum(10 .^( -1.0 .* dlvl ./ 10.0));

for k = 1 : n1

    atts = 10.^( -0.05 .* dlvl(k));

    if dlvl(k) >= 40.0
            atts = 0.0;
    end

    theta = th(k) .* pi ./ 180.0;

    [itmp,qtmp] = delay ( idata , qdata , nsamp ,
itau(k));
    [itmp3,qtmp3,ramp,rcos,rsin] = fade (itmp,qtmp,
nsamp,tstp,fd,n0(k),itn(k),flat);

  iout = iout + atts .* itmp3 ./ sqrt(total_attn);
  qout = qout + atts .* qtmp3 ./ sqrt(total_attn);

end
% *******************end of file********************

% ******************** STBC ********************

%About entering matrix O:
%-- O is Tp*Nt matrix,as default 4*3 complex
orthogonal is defined (rate 3/4).
%for [x1 -x2 -x3;x2* x1* 0;x3* 0 x1*;0 -x3* x2*]
-----enter------> O=[1 -2 -3;2+j 1+j 0;3+j 0
1+j;0 -3+j 2+j];
%-- Alamouti Scheme: [x1 x2;-x2* x1*]
-----enter------> O=[1 2;-2+j 1+j];
```

```
%-- A real orthogonal: [x1 x2;-x2 x1]
-----enter------> O=[1 2;-2 1];
%-- For real orthogonal matrices define M_psk=2; as
real signal constellation.
%-- O=[1]; is uncoded (no diversity).
%------------------------------------------------------------
clear all
O=[1 -2 -3;2+j 1+j 0;3+j 0 1+j;0 -3+j 2+j];
%Complex or Real Orthogonal Matrix **define this**
Nt=size(O,2);
%Number of Transmit antennas
co_time=size(O,1);
%Block time length
Nr=1;
%Number of Receive antennas        **define this**
Nit=100000;
%Number of repeats for each snr    **define this**
M_psk=4;
%M-PSK  constellation,M_psk=2^k    **define this**
snr_min=0;
%Min snr range for simulation      **define this**
snr_max=20;
%Max snr rande for simulation      **define this**
graph_inf_bit=zeros(snr_max-snr_min+1,2);
%Plot information
graph_inf_sym=zeros(snr_max-snr_min+1,2);
%Plot information
num_X=1;
num_bit_per_sym=log2(M_psk);
for cc_ro=1:co_time
    for cc_co=1:Nt
        num_X=max(num_X,abs(real(O(cc_ro,cc_co))));
    end
end
co_x=zeros(num_X,1);
for con_ro=1:co_time
%Compute delta,epsilon,eta and conj matrices
    for con_co=1:Nt
        if abs(real(O(con_ro,con_co)))~=0
delta(con_ro,abs(real(O(con_ro,con_co))))=
sign(real(O(con_ro,con_co)));
            epsilon(con_ro,abs(real(O(con_ro,
con_co))))=con_co;
co_x(abs(real(O(con_ro,con_co))),1)=
co_x(abs(real(O(con_ro,con_co))),1)+1;
```

```
eta(abs(real(O(con_ro,con_co)))),co_x(abs(real(O(con_
ro,con_co)))),1))=con_ro;
coj_mt(con_ro,abs(real(O(con_ro,con_co))))=imag(O(con_
ro,con_co));
        end
    end
end
eta=eta.';
%Sort is not necessary
eta=sort(eta);
eta=eta.';
for SNR=snr_min:snr_max
%Start simulation
    clc
    disp('Wait until SNR=');disp(snr_max);
    SNR
    n_err_sym=0;
    n_err_bit=0;
    graph_inf_sym(SNR-snr_min+1,1)=SNR;
    graph_inf_bit(SNR-snr_min+1,1)=SNR;
    for con_sym=1:Nit
        bi_data=randint(num_X,num_bit_per_sym);
%Random binary data
        de_data=bi2de(bi_data);
%Convert binary data to decimal for use in M-PSK mod.
        data=pskmod(de_data,M_psk,0,'gray');
        H=randn(Nt,Nr)+j*randn(Nt,Nr);
%Path gains matrix
        XX=zeros(co_time,Nt);
        for con_r=1:co_time
%Start space time coding
            for con_c=1:Nt
                if abs(real(O(con_r,con_c)))~=0
                    if imag(O(con_r,con_c))==0
XX(con_r,con_c)=data(abs(real(O(con_r,con_c))),1)*sign
(real(O(con_r,con_c)));
                    else
XX(con_r,con_c)=conj(data(abs(real(O(con_r,con_c))),1))
*sign(real(O(con_r,con_c)));
                    end
                end
            end
        end
%End space time coding
        H=H.';
```

```
        XX=XX.';
        snr=10^(SNR/10);
        Noise=(randn(Nr,co_time)+j*randn(Nr,co_time));
%Generate Noise
        Y=(sqrt(snr/Nt)*H*XX+Noise).';
%Received signal
        H=H.';
%Start decoding with perfect channel estimation
        for co_ii=1:num_X
            for co_tt=1:size(eta,2)
                if eta(co_ii,co_tt)~=0
                    if coj_mt(eta(co_ii,co_tt),co_ii)==0
r_til(eta(co_ii,co_tt),:,co_ii)=Y(eta(co_ii,co_tt),:);
a_til(eta(co_ii,co_tt),:,co_ii)=conj(H(epsilon(eta
(co_ii,co_tt),co_ii),:));
                    else
r_til(eta(co_ii,co_tt),:,co_ii)=conj(Y(eta(co_ii,
co_tt),:));
a_til(eta(co_ii,co_tt),:,co_ii)=H(epsilon(eta(co_
ii,co_tt),co_ii),:);
                    end
                end
            end
        end
        RR=zeros(num_X,1);
        for ii=1:num_X
%Generate decision statistics for the transmitted
signal "xi"
            for tt=1:size(eta,2)
                for jj=1:Nr
                    if eta(ii,tt)~=0
RR(ii,1)=RR(ii,1)+r_til(eta(ii,tt),jj,ii)*a_til(eta
(ii,tt),jj,ii)*delta(eta(ii,tt),ii);
                    end
                end
            end
        end
        re_met_sym=pskdemod(RR,M_psk,0,'gray');
% = ML decision for M-PSK
        re_met_bit=de2bi(re_met_sym);
        re_met_bit(1,num_bit_per_sym+1)=0;
%For correct demension of "re_met_bit"
        for con_dec_ro=1:num_X
            if re_met_sym(con_dec_ro,1)~=de_
data(con_dec_ro,1)
```

```
                    n_err_sym=n_err_sym+1;
                    for con_dec_co=1:num_bit_per_sym
                        if re_met_bit
(con_dec_ro,con_dec_co)~=bi_data(con_dec_ro,con_dec_co)
                            n_err_bit=n_err_bit+1;
                        end
                    end
                end
            end
        end
        Perr_sym=n_err_sym/(num_X*Nit);
%Count number of error bits and symbols
        graph_inf_sym(SNR-snr_min+1,2)=Perr_sym;
        Perr_bit=n_err_bit/(num_X*Nit*num_bit_per_sym);
        graph_inf_bit(SNR-snr_min+1,2)=Perr_bit;
end
x_sym=graph_inf_sym(:,1);
%Generate plot
y_sym=graph_inf_sym(:,2);
subplot(2,1,1);
semilogy(x_sym,y_sym,'k-v');
xlabel('SNR, [dB]');
ylabel('Symbol Error Probability');
grid on
x_bit=graph_inf_bit(:,1);
y_bit=graph_inf_bit(:,2);
subplot(2,1,2);
semilogy(x_bit,y_bit,'k-v');
xlabel('SNR, [dB]');
ylabel('Bit Error Probability');
grid on

% ****************************************************
%%%%%%%%%%%%%%%%%%Conventional SLM Scheme %%%%%%%%%%%%%%
% ****************************************************

clc
clear all

nFFT = 128; % fft size
nDSC = 114; % number of data subcarriers
nSym = 10^4; % number of symbols
g=0;
papr_dB_min=zeros(1,nSym);
```

```
for n =1:nSym; % number of symbols
    g=g+1;
x=randsrc(1,nDSC*nSym,[-1 1 ]); % data
% phase sequences
B_1 = ones(1,nDSC*nSym);
B_2 = randsrc(1,nDSC*nSym);
B_3 = randsrc(1,nDSC*nSym);
B_4 = randsrc(1,nDSC*nSym);


        X = x.*B_1;
        Y = reshape(X,nDSC,nSym).'; % grouping into
multiple symbols
        % Assigning modulated symbols to subcarriers
        Y = [zeros(nSym,6) Y(:,[1:nDSC/2])
zeros(nSym,3) Y(:,[nDSC/2+1:nDSC]) zeros(nSym,5)] ;
        y = (nFFT/sqrt(nDSC))*ifft(fftshift(Y.')).';%
Taking iFFT, time domain
        meanSquareValue = sum(y.*conj(y),2)/nFFT; %
computing the peak to average power ratio
        peakValue = max(y.*conj(y),[],2);
        papr = peakValue ./meanSquareValue;
        papr_dB_1= 10*log10(papr);


        X= x.*B_2;
        Y = reshape(X,nDSC,nSym).'; % grouping into
multiple symbols
        Y = [zeros(nSym,6) Y(:,[1:nDSC/2])
zeros(nSym,3) Y(:,[nDSC/2+1:nDSC]) zeros(nSym,5)] ;
        y = (nFFT/sqrt(nDSC))*ifft(fftshift(Y.')).';%
Taking iFFT, time domain
        meanSquareValue = sum(y.*conj(y),2)/nFFT;
        peakValue = max(y.*conj(y),[],2);
        papr = peakValue ./meanSquareValue;
        papr_dB_2= 10*log10(papr);


        X= x.*B_3;
        Y = reshape(X,nDSC,nSym).'; % grouping into
multiple symbols
        Y = [zeros(nSym,6) Y(:,[1:nDSC/2])
zeros(nSym,3) Y(:,[nDSC/2+1:nDSC]) zeros(nSym,5)] ;
        y = (nFFT/sqrt(nDSC))*ifft(fftshift(Y.')).';%
Taking iFFT, time domain
        meanSquareValue = sum(y.*conj(y),2)/nFFT;
        peakValue = max(y.*conj(y),[],2);
```

```
        papr = peakValue ./meanSquareValue;
        papr_dB_3= 10*log10(papr);


        X= x.*B_4;
        Y = reshape(X,nDSC,nSym).'; % grouping into
multiple symbols
        Y = [zeros(nSym,6) Y(:,[1:nDSC/2])
zeros(nSym,3) Y(:,[nDSC/2+1:nDSC]) zeros(nSym,5)] ;
        y = (nFFT/sqrt(nDSC))*ifft(fftshift(Y.')).';%
Taking iFFT, time domain
        meanSquareValue = sum(y.*conj(y),2)/nFFT;
        peakValue = max(y.*conj(y),[],2);
        papr = peakValue ./meanSquareValue;
        papr_dB_4= 10*log10(papr);


papr_dB=[papr_dB_1 papr_dB_2 papr_dB_3 papr_dB_4];
end
m=0;
for j=1:nSym
    m=m+1;
[papr_dB_min(1,m),si]=min(papr_dB(m,:));
SI(m)=si;
end
papr_plot=papr_dB_min;
[nBPSK1 xBPSK1] = hist(papr_plot,[0:0.5:15]);
semilogy(xBPSK1,1-cumsum(nBPSK1)/nSym,'bs-')
xlabel('papr, x dB')
ylabel('Probability, PAPR > x')
hold on
grid on
%axis([3 12 0 1])


% ****************************************************
%%%%%%%%%%%%%%%       SISO_SLM system    %%%%%%%%%%%%%%
% ****************************************************
clear all
K=128;
alph=1;
K1=K*alph;
V=4;
i=0;
for R=1:0.5:15
    i=i+1;
    P(i)=(1-(1-exp(-R))^K1)^V;
```

```
end
R=1:0.5:15;
Rd=10*log10(R);
semilogy(Rd,P,'b')
axis([5 12 1e-4 1e0])
grid on
hold on
xlabel('PAPRo [dB]')
ylabel('Prob (PAPRlow > PAPRo )')


% *******************************************************
%%%%%%%%%% Simplified system (MIMO d-SLM) %%%%%%%%%%%
% *******************************************************
clear all
K=128;
alph=1;
K1=K*alph;
V=4;
M=2;
delta=10^(0.45/10);
i=0;
for Rd=0:0.5:20
    i=i+1;
    R=10^(Rd/10);
    P(i)=(1-(1-exp(-R/delta))^(K1))^(M*V);
end
Rd=0:0.5:20;
%Rd=10*log10(R);
semilogy(Rd,P,'b')
axis([5 12 1e-4 1e0])
grid on
hold on


% *******************************************************
%%%%%%%%%% Simplified system (MIMO s-SLM)%%%%%%%%%%%%
% *******************************************************

clear all
K=128;
alph=1;
K1=K*alph;
V=4;
M=2;
i=0;
```

```
for R=2:0.5:9
    i=i+1;
    P(i)=(1-(1-exp(-R))^(M*K1))^V;
end
R=2:0.5:9;
Rd=10*log10(R);
semilogy(Rd,P,'b')
axis([5 12 1e-4 1e0])
grid on
hold on

% ****************************************************
%%%%%% detection Prob. for SISO and MIMO-iSLM %%%%%%
% ****************************************************

% for SISO ---> N=1, M=1;snr=10.^(snrd/10);
% for MIMO-iSLM ---> N & M not equal one;

clear all
N=1;% no. of receiving ant.
%M=2;
N_SI=2;% SI bits according to the type of SLM
V=4;
l=0;
for snrd=0:1:20
    l=l+1;
    snr=10^(snrd/10);
    lam=(1+snr^-1)^-0.5;
    sum0=0;
    for n=0:N
        sum0=sum0+(factorial(N-1+n)./(factorial(n).
*factorial(N-1))).*((1+lam)/2).^n;
        Pb=(((1-lam)/2)^N).*sum0;
    %Pd(l)=1-(M*ceil((log10(V))/log10(2))*Pb);
    Pd(l)=1-(N_SI*Pb);
    end
end
snrd=0:1:20;
%snrd=10*log10(snr);
plot(snrd,Pd,'k')
grid on
xlabel('SNR [dB]')
ylabel('Prob. of SI Bits Detection ')
axis([0 20 0 1])
hold on
```

```
% ********************************************************
%%%%%%%    Overall BER for SISO and MIMO-iSLM    %%%%%%%
% ********************************************************
% for SISO ---> N=1, M=1;
% for MIMO-iSLM ---> N & M not equal one;

clear all
N=2;
N_SI=2;% SI bits according to the type of SLM
V=4;
l=0;
for snrd=0:1:20
    l=l+1;
    snr=10^(snrd/10);
    lam=(1+snr^-1)^-0.5;
    sum0=0;
    for n=0:N
        sum0=sum0+(factorial(N-1+n)./(factorial(n).
*factorial(N-1))).*((1+lam)/2).^n;
        Pb=(((1-lam)/2)^N).*sum0;
    %Pd(l)=1-(M*ceil((log10(V))/log10(2))*Pb);
     Pd(l)=1-(N_SI*Pb);
    P0(l)=(Pb.*Pd(l))+((1-Pd(l))/2);
end
end
snrd=0:1:20;
%snrd=10*log10(snr);
%ebn0=snrd-3;
semilogy(snrd,P0,'r')
grid on
xlabel('SNR [dB]')
ylabel('Overall Bit Error Rate ')
axis([0 20 1e-4 1])
hold on

% ********************************************************
%%%%%%%        PAPR for %unequal power dist.     %%%%%%%
% ********************************************************
% PAPR for unmodified data
clear all

nFFT = 128; % fft size
nDSC = 114; % number of data subcarriers
nSym = 5*10^4; % number of symbols
```

```
ipQPSK = 1/sqrt(2)*(randsrc(1,nDSC*nSym,[-1 1 ]) +
j*randsrc(1,nDSC*nSym,[-1 1 ]));

ipModQPSK = reshape(ipQPSK,nDSC,nSym).'; % grouping
into multiple symbols

% Assigning modulated symbols to subcarriers

xQPSK1F = [zeros(nSym,6) ipModQPSK(:,[1:nDSC/2])
zeros(nSym,3)... %unequal power dist.
 ipModQPSK(:,[nDSC/2+1:nDSC]) zeros(nSym,5)] ;

% Taking iFFT, time domain
xQPSK1t = (nFFT/sqrt(nDSC))*ifft(fftshift(xQPSK1F.')).';

% computing the peak to average power ratio

meanSquareValueQPSK1 = sum(xQPSK1t.*conj(xQPSK1t),2)/
nFFT;
peakValueQPSK1 = max(xQPSK1t.*conj(xQPSK1t),[],2);
paprSymbolQPSK1 = peakValueQPSK1./
meanSquareValueQPSK1;
paprSymbolQPSK1dB = 10*log10(paprSymbolQPSK1);
[nQPSK1 xQPSK1] = hist(paprSymbolQPSK1dB,[0:0.5:15]);

semilogy(xQPSK1,1-cumsum(nQPSK1)/nSym,'ro-')

xlabel('papr, x dB')
ylabel('Probability, PAPR > x')
hold on
grid on
%axis([3 12 0 1])
```

Appendix B: MATLAB® Simulation Codes for Chapters 5 through 8

```
%*****************************************************
% ----------- Simulation program for CPM-OFDM -----------
%*****************************************************

clear all

Trans_max=100e6; % max bits sent per SNR
Trans_min=1e6; % min bits sent per SNR
Error_min=2e5; % min errors per SNR
%********************%%%%++++++%%%%********************
targetBER=1e-5; % target BER
SNRmax=50; % max SNR (dB)
Ndft=512; % DFT size (for equalizer)
J=8; % oversampling factor                           <-----
---------- change -------------*
N=64; % number of subcarriers ---> (N = Ndft/J)
io=1; % index offset
A=1; % signal amplitude
M=4; % modulation order =====> sqrt(Mqam = 4, 16,
64, 256) <-------- change ----------*
modh=1/(2*pi); % modulation index                    <------
--------- change -------------*
```

```
TB=128e-6; % block time
Tg=10e-6;  % guard time
TF=Tg+TB;  % frame time (Tr. efficiency = TB/TF)
Fsa=J*N/TB; % sampling rate
Tsa=1/Fsa; % sampling period
Ng=Tg*Fsa; % samples per guard interval
NB=TB*Fsa; % samples per symbol
NF=TF*Fsa; % samples per frame
ip=[Ng:NF-1]+io; % processing indices for s(t)
taumax=9e-6; % maximum delay spread of channel
(sec)<---- change
Nc=taumax*Fsa; % number of channel taps
Nr=Nc+NF-1; % number of received samples
L=8; % blocks/channel realization (vectorize)paths

%*** Bit and Symbol mappings (depends on modulation
order)******

if M==2
SymMap=[-1;1]; % data symbol mapping
BitMap=[0; 1]; % bit mapping
end
if M==4
SymMap=[-3;-1;1;3]; % data symbol mapping
BitMap=[... % bit mapping
0 0; 0 1; 1 1; 1 0];
end
if M==8
SymMap=[-7:2:7]'; % data symbol mapping
BitMap=[... % bit mapping
0 0 0; 0 0 1; 0 1 1; 0 1 0; 1 1 0; 1 1 1; 1 0 1; 1
0 0];
end
if M==16
SymMap=(-15:2:15)'; % data symbol mapping
BitMap=[... % bit mapping
0 0 0 0; 0 0 0 1; 0 0 1 1; 0 0 1 0; 0 1 1 0; 0 1 1 1;
...
0 1 0 1; 0 1 0 0; 1 1 0 0; 1 1 0 1; 1 1 1 1; 1 1 1 0;
...
1 0 1 0; 1 0 1 1; 1 0 0 1; 1 0 0 0];
end
varI=sum(SymMap.^2)/M; % variance of data (M^2-1)/3 if
data is iid
CN=sqrt(2/(N*varI)); % normalizing constant
```

```
%**************   Subcarrier Matrix   ***************

t=0:Tsa:(TB-Tsa); % time vector
W=zeros(NB,N); % initialize unitary matrix
for k=1:N/2 % W is a set of orth. sines and
cosines....eq(4)
W(:,k)=cos(2*pi*k*t/TB)';
end
for k=(N/2+1):N
W(:,k)=sin(2*pi*(k-N/2)*t/TB)';
end
%
% Design FIR filter to improve performance of phase
demodulator

Mf=11; % filter length 3=< Lfilter<=101
n1=0:(Mf-1); % filter sample index
d=(Mf-1)/2; % delay
n2=(d+1):(d+NB); % desired, delayed indices
fc=0.2; % normalized cutoff frequency (cyc/samp)--->
0:1
wc=2*pi*fc; % normalized cutoff frequency (rad/samp)
h1=zeros(1,Mf); % initialize
for i=1:Mf % compute coefficients
if n1(i)==((Mf-1)/2)
h1(i)=wc/pi;
else
h1(i)=sin(wc*(n1(i)-(Mf-1)/2))/(pi*(n1(i)-(Mf-1)/2));
end
end
w1=0.54-0.46*cos(2*pi*n1/(Mf-1)); % Hamming window
hf=h1.*w1; % windowed filter coefficients


                %<----- change (A, B, C, and D) ----*
%%%%%%%%%%%%%%%%%%%%%%%%%%%%%%%%%%%%%%%%%%%%%%%%%%%%%%%%%%
%**************** Channel A (two-path)****************

tau=[0 5e-6]; % path delays
power_dB=[0 -10]; % path power (dB)
power=10.^(power_dB/10); % path power
for n=1:length(tau)
i=tau(n)*Fsa; % path index
p(i+io,1)=power(n); % delay PSD
end
p=[p; zeros(Nc-length(p),1)]; % zero-pad
```

```
%******************************************************
```

```
%***************** Channel B (two-path)***************
```

```
%tau=[0 5e-6]; % path delays
%power_dB=[0 -3]; % path power (dB)
%power=10.^(power_dB/10); % path power
%for n=1:length(tau)
%i=tau(n)*Fsa; % path index
%p(i+io,1)=power(n); % delay PSD
%end
%p=[p; zeros(Nc-length(p),1)]; % zero-pad
```

```
%******************************************************
```

```
%*************** Channel C (exponential)*************
```

```
%tau=[0:Nc-1]'*Tsa; % time vector
%p=(1/taumax)*exp(-tau/2e-6); % delay PSD
```

```
%******************************************************
```

```
%***************** Channel D (uniform)***************
```

```
%tau=[0:Nc-1]'*Tsa; % discrete propagation delays
%p=ones(size(tau)); % delay PSD
```

```
%%%%%%%%%%%%%%%%%%%%%%%%%%%%%%%%%%%%%%%%%%%%%%%%%%%%%%%
```

```
% ---------------- Simulation loop -----------------
```

```
BER=0; % initialize BER vector
EbN0_dB=0; % initialize SNR vector
dx=2.5; % SNR step size
iSNR=1; % SNR counter
go=1; % initialize loop
while go % run until max SNR condition
Error_num=0; Trans_num=0; % initialize
while Trans_num<=Trans_min | ...
(Error_num<=Error_min & Trans_num<=Trans_max)
```

```
%********   Generate L blocks; phi; theta0   ********
```

```
in=ceil(M*rand(N,L)); % random symbol index
I=SymMap(in); % data symbols
```

```
m=CN*W*I; % OFDM message signal
theta0=2*pi*rand(1,L)-pi; % memory terms (assume
uniform)
phi=zeros(NF,L); % initialize CPM-OFDM phase signal
for i=1:L % cyclic prefix
phi(:,i)=[2*pi*modh*m(NB-Ng+1:NB,i)+theta0(i);2*pi*modh*m
(:,i)+theta0(i)];
end

s=A*exp(j*phi); % CPM-OFDM signal

%*************   Determine noise power   **************

Es=sum(sum(abs(s).^2))*Tsa; % signal energy
Eb=Es/(L*N*log2(M)); % bit energy
EbN0=10^(EbN0_dB(iSNR)/10); % SNR
N0=Eb./EbN0; % noise spectral height

%%%%*****************   Channel    *****************

tmp=sqrt(1/2)*(randn(Nc,1)+j*randn(Nc,1)); % Gaussian
vector
Ch=sqrt(p/sum(p)).*tmp; % channel (normalize average
power)

% Received signal plus noise (to be processed by FDE)
********

rp=zeros(NB,L); % initialize
for i=1:L
tmp1=(conv(Ch,s(:,i))).'; % received samples
tmp1=tmp1(ip); % discard cyclic prefix
tmp2=sqrt(1/2)*(randn(NB,1)+j*randn(NB,1)); % complex
Gaussian
noise=sqrt(N0*Fsa)*tmp2; % Gaussian noise
rp(:,i)=tmp1+noise'; % received samples plus noise
end

%%%*******  Frequency-domain equalizer (FDE) **********

H=fft(Ch,Ndft); % channel gains
C=conj(H)./(abs(H).^2+EbN0^(-1)); % correction term (MMSE)
X=fft(rp,Ndft); % to frequency domain
hatS=X.*(C*ones(1,L)); % equalize
x=ifft(hatS,Ndft); % to time domain
```

```
%%%************   Filter signal   ****************

hats=zeros(NB,L); % initialize
for i=1:L
tmp=(conv(hf,x(:,i))).'; % filtered signal
hats(:,i)=tmp(n2); % filtered signal, desired indices
end

%%************%   Demodulate and detect   ************

hatphi=unwrap(angle(hats)); % phase demodulate
Ihat=W'*hatphi/((2*pi*modh*CN)*NB*1/2); % matched-
filter output
inHat=min(round((Ihat+(M-1))/2)+io,M); % index
estimate, (<=M)
inHat=max(inHat,1); % (>=1)
Errors=sum(sum(BitMap(in,:)~=BitMap(inHat,:))); % bit
errors
Error_num=Error_num+Errors; % cumulative bit errors
Trans_num=Trans_num+L*N*log2(M); % cumulative bits

end % end this SNR

BER(iSNR)=Error_num/Trans_num; % bit error rate for
current SNR

%%% Test for max SNR condition

if BER(iSNR)<targetBER | EbN0_dB(iSNR)>=SNRmax
go=0;
else % keep going
iSNR=iSNR+1;
EbN0_dB(iSNR)=EbN0_dB(iSNR-1)+dx;
end

end %***************% end simulation %****************

%%%******************   Plot   ********************

semilogy(EbN0_dB,BER,'r')
grid on
axis([0 50 1e-4 1e0])
xlabel('Eb/No [dB]')
ylabel('Bit Error Rate')
```

```
%*****************************************************
% ---Simulation program for CPM-OFDM based chaotic---%
%--------------    interleaving scheme    ------------%
%*****************************************************

clear all

Trans_max=10e6; % max bits sent per SNR
Trans_min=1e6; % min bits sent per SNR
Error_min=2e5; % min errors per SNR

targetBER=1e-5; % target BER
SNRmax=50; % max SNR (dB)
Ndft=512; % DFT size (for equalizer)
J=8; % oversampling factor                    <-----
---------- change -------------*
N=64; % number of subcarriers ---> (N = Ndft/J)
io=1; % index offset
A=1; % signal amplitude
M=4; % modulation order =====> sqrt(Mqam = 4, 16,
64, 256) <-------- change ----------*
modh=1/(2*pi); % modulation index             <------
--------- change -------------*
TB=128e-6; % block time
Tg=10e-6;  % guard time
TF=Tg+TB;  % frame time (Tr. efficiency = TB/TF)
Fsa=J*N/TB; % sampling rate
Tsa=1/Fsa; % sampling period
Ng=Tg*Fsa; % samples per guard interval
NB=TB*Fsa; % samples per symbol
NF=TF*Fsa; % samples per frame
ip=[Ng:NF-1]+io; % processing indices for s(t)
taumax=9e-6; % maximum delay spread of channel (sec)<-
---------- change -------------*
Nc=taumax*Fsa;%taumax*Fsa; % number of channel taps
Nr=Nc+NF-1; % number of received samples
L=8; % blocks/channel realization (vectorize)paths
%alpha=2;
% Bit and Symbol mappings (depends on modulation order)
******

if M==2
SymMap=[-1;1]; % data symbol mapping
BitMap=[0; 1]; % bit mapping
```

```
end
if M==4
SymMap=[-3;-1;1;3]; % data symbol mapping
BitMap=[... % bit mapping
0 0; 0 1; 1 1; 1 0];
end
if M==8
SymMap=[-7:2:7]'; % data symbol mapping
BitMap=[... % bit mapping
0 0 0; 0 0 1; 0 1 1; 0 1 0; 1 1 0; 1 1 1; 1 0 1; 1
0 0];
end
if M==16
SymMap=(-15:2:15)'; % data symbol mapping
BitMap=[... % bit mapping
0 0 0 0; 0 0 0 1; 0 0 1 1; 0 0 1 0; 0 1 1 0; 0 1 1 1;
...
0 1 0 1; 0 1 0 0; 1 1 0 0; 1 1 0 1; 1 1 1 1; 1 1 1 0;
...
1 0 1 0; 1 0 1 1; 1 0 0 1; 1 0 0 0];
end
varI=sum(SymMap.^2)/M; % variance of data (M^2-1)/3 if
data is iid
CN=sqrt(2/(N*varI)); % normalizing constant

%**************** Subcarrier Matrix*****************

t=0:Tsa:(TB-Tsa); % time vector
W=zeros(NB,N); % initialize unitary matrix
for k=1:N/2 % W is a set of orth. sines and
cosines....eq(4)
W(:,k)=cos(2*pi*k*t/TB)';
end
for k=(N/2+1):N
W(:,k)=sin(2*pi*(k-N/2)*t/TB)';
end
%
%Design FIR filter to improve performance of phase
demodulator

Mf=11; % filter length 3=< Lfilter<=101
n1=0:(Mf-1); % filter sample index
d=(Mf-1)/2; % delay
n2=(d+1):(d+NB); % desired, delayed indices
```

```
fc=0.2; % normalized cutoff frequency (cyc/samp)---> 0:1
wc=2*pi*fc; % normalized cutoff frequency (rad/samp)
h1=zeros(1,Mf); % initialize
for i=1:Mf % compute coefficients
if n1(i)==((Mf-1)/2)
h1(i)=wc/pi;
else
h1(i)=sin(wc*(n1(i)-(Mf-1)/2))/(pi*(n1(i)-(Mf-1)/2));
end
end
w1=0.54-0.46*cos(2*pi*n1/(Mf-1)); % Hamming window
hf=h1.*w1; % windowed filter coefficients

% Channel delay power spectral density(C:exponential)
                %<----- change (A, B, C, and D) ----*
tau=[0:Nc-1]'*Tsa; % time vector
p=0.1188*exp(-tau/2e-6); % delay PDS

% ------------------ Simulation loop ------------------

BER=0; % initialize BER vector
EbN0_dB=0; % initialize SNR vector
dx=2.5; % SNR step size
iSNR=1; % SNR counter
go=1; % initialize loop
while go % run until max SNR condition
Error_num=0; Trans_num=0; % initialize
while Trans_num<=Trans_min | ...
(Error_num<=Error_min & Trans_num<=Trans_max)

%********   Generate L blocks; phi; theta0   ********

in=ceil(M*rand(N,L)); % random symbol index
I=SymMap(in); % data symbols
m=CN*W*I; % OFDM message signal
%
theta0=2*pi*rand(1,L)-pi; % memory terms (assume uniform)
phi=zeros(NF,L); % initialize CPM-OFDM phase signal
for i=1:L % cyclic prefix
phi(:,i)=[2*pi*modh*m(NB-Ng+1:NB,i)+theta0(i);...
2*pi*modh*m(:,i)+theta0(i)];
end

s=A*exp(j*phi); % CPM-OFDM signal (transmitted signal)
```

```
%&&&&&&&&&&&&&&& Chaotic Interleaving &&&&&&&&&&&&&&

ss1=s(1:40,:);
ss2=s(41:end,:);
[ss2,dim1,dim2]=randomize(ss2);
ss1=ss2(NB-Ng+1:NB,:);% cyclic prefix
s=[ss1;ss2];
%*************   Determine noise power   *************

Es=sum(sum(abs(s).^2))*Tsa; % signal energy
Eb=Es/(L*N*log2(M)); % bit energy
EbN0=10^(EbN0_dB(iSNR)/10); % SNR
N0=Eb./EbN0; % noise spectral height

%%%%*****************    Channel    ****************

tmp=sqrt(1/2)*(randn(Nc,1)+j*randn(Nc,1)); % Gaussian
vector
Ch=sqrt(p/sum(p)).*tmp; % channel (normalize average
power)

%%% Received signal plus noise (to be processed by FDE)
********

rp=zeros(NB,L); % initialize
for i=1:L
tmp1=(conv(Ch,s(:,i))).'; % received samples
tmp1=tmp1(ip); % discard cyclic prefix
tmp2=sqrt(1/2)*(randn(NB,1)+j*randn(NB,1)); % complex
Gaussian
noise=sqrt(N0*Fsa)*tmp2; % Gaussian noise
rp(:,i)=tmp1+noise'; % received samples plus noise
%[XD,CXD,LXD] = wden(rp,'sqtwolog','s','one',2,
'sym8');
%rp=reshape(XD,512,8);
end

%%%*******  Frequency-domain equalizer (FDE) *********

H=fft(Ch,Ndft); % channel gains
C=conj(H)./(abs(H).^2+EbN0^(-1)); % correction term
(MMSE)1./H;%
X=fft(rp,Ndft); % to frequency domain
hatS=X.*(C*ones(1,L)); % equalize
x=ifft(hatS,Ndft); % to time domain
```

```
%&&&&&&&&&&&&& Chaotic De-Interleaving &&&&&&&&&&&&&&&
x=derandomize(x);
%%%***************    Filter signal    ***************

hats=zeros(NB,L); % initialize
for i=1:L
tmp=(conv(hf,x(:,i))).'; % filtered signal
hats(:,i)=tmp(n2); % filtered signal, desired indices
end

%%************%    Demodulate and detect   *************

hatphi=unwrap(angle(hats)); % phase demodulate

Ihat=W'*hatphi/((2*pi*modh*CN)*NB*1/2); % matched-
filter output
inHat=min(round(((Ihat+(M-1))/2)+io,M); % index
estimate, (<=M)
inHat=max(inHat,1); % (>=1)
Errors=sum(sum(BitMap(in,:)~=BitMap(inHat,:))); % bit
errors
Error_num=Error_num+Errors; % cumulative bit errors
Trans_num=Trans_num+L*N*log2(M); % cumulative bits

end % end this SNR

BER(iSNR)=Error_num/Trans_num; % bit error rate for
current SNR

%%% .Test for max SNR condition

if BER(iSNR)<targetBER | EbN0_dB(iSNR)>=SNRmax
go=0;
else % keep going
iSNR=iSNR+1;
EbN0_dB(iSNR)=EbN0_dB(iSNR-1)+dx;
end

end %*************% end simulation %*****************

%%%********************    Plot    ******************

semilogy(EbN0_dB,BER,'r-o')
grid on
axis([0 50 1e-4 1e0])
```

```
xlabel('Eb/No [dB]')
ylabel('Bit Error Rate')
hold on

%**********************************************************
function [y,m,n]=randomize(x)
[m,n]=size(x);
y=reshape(x,sqrt(m*n),sqrt(m*n));
n1 = [sqrt(n*m)/8,sqrt(n*m)/8,sqrt(n*m)/8,sqrt(n*m)/
8,sqrt(n*m)/8,sqrt(n*m)/8,sqrt(n*m)/8,sqrt(n*m)/8];
%n=[256,256]
[pr,pc] = chaomat(n1);
y = chaoperm(y,pr,pc,3,'forward');
y=reshape(y,m,n);


%**********************************************************

function x=derandomize(y)
[m,n]=size(y);
x=reshape(y,sqrt(m*n),sqrt(m*n));
n1 = [sqrt(n*m)/8,sqrt(n*m)/8,sqrt(n*m)/8,sqrt(n*m)/
8,sqrt(n*m)/8,sqrt(n*m)/8,sqrt(n*m)/8,sqrt(n*m)/8];
[pr,pc] = chaomat(n1);

x = chaoperm(x,pr,pc,3,'backward');
x=reshape(x,m,n);

%**********************************************************

function [pr,pc]=chaomat(n)
%
I=sum(n);
k=size(n,2);
for i=1:k
    N(i+1)=1;
    for j=1:i
        N(i+1)=N(i+1)+n(j);
    end
end
N(1)=1;
%N(0)=1;
for cb=1:k
   for rb=1:n(cb)
      rbstartcol(rb)=mod((rb-1)*I,n(cb));
      rbendcol(rb)=mod((rb*I-1),n(cb));
```

```
      rbstartrow(rb)=fix(((rb-1)*I)/n(cb));
      rbendrow(rb)=fix((rb*I-1)/n(cb));
      mincol(rb)=min([rbendcol(rb)+1,rbstartcol(rb)]);
      maxcol(rb)=max([rbendcol(rb),rbstartcol(rb)-1]);
   end

   for i=1:I
      for j=N(cb):N(cb+1)-1
         newindex(i,j-N(cb)+1)=(i-1)*n(cb)+(n(cb)
-j+N(cb)-1);
         newindexmod(i,j-N(cb)+1)=mod(newindex
(i,j-N(cb)+1),n(cb));
         newindexquotient(i,j-N(cb)+1)=fix(newindex
(i,j-N(cb)+1)/n(cb));
         rowblockindex(i,j-N(cb)+1)=fix(newindex
(i,j-N(cb)+1)/I)+1;
      end
   end

   for i=1:I
      for j=1:n(cb)
         for rb=1:n(cb)
            if rowblockindex(i,j)==rb;
               if newindexmod(i,j)>maxcol(rb)
                  col=rbendrow(rb)-newindexquotient
(i,j)+(n(cb)-1-newindexmod(i,j))*(rbendrow(rb)-
rbstartrow(rb));
               elseif newindexmod(i,j)>=mincol(rb) &
newindexmod(i,j)<=maxcol(rb)
                  if rbstartcol(rb)>rbendcol(rb)
                     c=0;
                     d=-1;
                  else
                     c=1;
                     d=1;
                  end
                  col=(rbendrow(rb)-rbstartrow(rb))
*(n(cb)-1-maxcol(rb))+(rbendrow(rb)-newindexquotient
(i,j)+c)+(maxcol(rb)-newindexmod(i,j))*(rbendrow(rb)-
rbstartrow(rb)+d);
               else %if newindexmod(i,j)<=mincol(rb)
                  col=I-mincol(rb)*(rbendrow(rb)-
rbstartrow(rb))+(rbendrow(rb)-newindexquotient(i,j)+1)
+(mincol(rb)-1-newindexmod(i,j))*(rbendrow(rb)-
rbstartrow(rb));
```

```
            end

                row=1+I-N(cb+1)+rowblockindex(i,j);
            end
        end

        pr(i,j+N(cb)-1)=row;
        pc(i,j+N(cb)-1)=col;
    end
  end
end

%*******************************************************
function out=chaoperm(im,pr,pc,num,forward)
%
[rows,cols] = size(im);
mat = zeros([rows,cols,num+1]);
mat(:,:,1) = im(:,:);

for loc=2:num+1
    if(strcmp(forward,'forward'))
        for i=1:rows
            for j=1:cols
                mat(pr(i,j),pc(i,j),loc) =
mat(i,j,loc-1);
            end
        end
    elseif(strcmp(forward,'backward'))
        for i=1:rows
            for j=1:cols
                mat(i,j,loc) = mat(pr(i,j),pc(i,j),
loc-1);
            end
        end
    end
end
out = mat(:,:,num+1);

%*******************************************************
function test()

im=imread('lenna.jpg');
im=double(im);
%imshow(im);
```

```
% the sum of n should be equal to the size of image
im. Here lenna is 512x512, i.e., sum(n)==512
n = [10,5,12,5,10,8,14,10,5,12,5,10,8,14,10,5,12,5,10,
8,14,10,5,12,5,10,8,14,10,5,12,5,10,8,14,10,5,12,5,10,
8,14,10,5,12,5,10,8,14,10,5,12,5,10,8,14,];
[pr,pc] = chaomat(n);
pim = chaoperm(im,pr,pc,3,'forward');

figure(1);
imshow(pim);

% restore the original image
rim = chaoperm(pim,pr,pc,3,'backward');

figure(2);
imshow(rim);

%*****************************************************
function out=chaoperm(im,pr,pc,num,forward)
%
[rows,cols] = size(im);
mat = zeros([rows,cols,num+1]);
mat(:,:,1) = im(:,:);

for loc=2:num+1
    if(strcmp(forward,'forward'))
        for i=1:rows
            for j=1:cols
                mat(pr(i,j),pc(i,j),loc) = mat(i,j,
loc-1);
            end
        end
    elseif(strcmp(forward,'backward'))
        for i=1:rows
            for j=1:cols
                mat(i,j,loc) = mat(pr(i,j),pc(i,j),
loc-1);
            end
        end
    end
end
out = mat(:,:,num+1);

function [pr,pc]=chaomat(n)
```

```
%
I=sum(n);
k=size(n,2);
for i=1:k
    N(i+1)=1;
    for j=1:i
        N(i+1)=N(i+1)+n(j);
    end
end
N(1)=1;
%N(0)=1;
for cb=1:k
    for rb=1:n(cb)
        rbstartcol(rb)=mod((rb-1)*I,n(cb));
        rbendcol(rb)=mod((rb*I-1),n(cb));
        rbstartrow(rb)=fix(((rb-1)*I)/n(cb));
        rbendrow(rb)=fix((rb*I-1)/n(cb));
        mincol(rb)=min([rbendcol(rb)+1,rbstartcol(rb)]);
        maxcol(rb)=max([rbendcol(rb),rbstartcol(rb)-1]);
    end

    for i=1:I
        for j=N(cb):N(cb+1)-1
            newindex(i,j-N(cb)+1)=(i-1)*n(cb)+
(n(cb)-j+N(cb)-1);
            newindexmod(i,j-N(cb)+1)=
mod(newindex(i,j-N(cb)+1),n(cb));
            newindexquotient(i,j-N(cb)+1)=
fix(newindex(i,j-N(cb)+1)/n(cb));
            rowblockindex(i,j-N(cb)+1)=
fix(newindex(i,j-N(cb)+1)/I)+1;
        end
    end

    for i=1:I
        for j=1:n(cb)
            for rb=1:n(cb)
                if rowblockindex(i,j)==rb;
                    if newindexmod(i,j)>maxcol(rb)
                        col=rbendrow(rb)-newindexquotient(i,j)
+(n(cb)-1-newindexmod(i,j))*(rbendrow(rb)-
rbstartrow(rb));
                    elseif newindexmod(i,j)>=mincol(rb) &
newindexmod(i,j)<=maxcol(rb)
                        if rbstartcol(rb)>rbendcol(rb)
```

```
                        c=0;
                        d=-1;
                    else
                        c=1;
                        d=1;
                    end
                    col=(rbendrow(rb)-
rbstartrow(rb))*(n(cb)-1-maxcol(rb))+(rbendrow(rb)-new
indexquotient(i,j)+c)+(maxcol(rb)-
newindexmod(i,j))*(rbendrow(rb)-rbstartrow(rb)+d);
                    else %if newindexmod(i,j)<=mincol(rb)
                        col=I-mincol(rb)*(rbendrow(rb)-
rbstartrow(rb))+(rbendrow(rb)-newindexquotient(i,j)+1)
+(mincol(rb)-1-newindexmod(i,j))*(rbendrow(rb)-
rbstartrow(rb));
                    end

                    row=1+I-N(cb+1)+rowblockindex(i,j);
                end
            end

            pr(i,j+N(cb)-1)=row;
            pc(i,j+N(cb)-1)=col;
        end
    end
end


%*****************************************************
            %Image Transmission over SC-FDMA
%*****************************************************

clear all
tic
%======= Choose simulation Parameters
SP.FFTsize = 256;
SP.inputBlockSize = 32;
SP.CPsize = 20;
%SP.subband = 15;
SP.subband = 0;

SP.SNR = 30;

%======= Choose channel type

%======= Uniform Channel
```

```
%         SP.channel1=(1/sqrt(10))*(randn(10,SP.
numRun)+sqrt(-%1)*randn(10,SP.numRun))/sqrt(2);


%======= SUI3 Channel

SP.paths= fadchan(SP);

%======= Vehicular A Channel

% x(1)=1;
% x(2)=10^(-1/10);
% x(3)=10^(-9/10);
% x(4)=10^(-10/10);
% x(5)=10^(-15/10);
% x(6)=10^(-20/10);
% tuamp=sqrt(x);
% tp=tuamp/norm(tuamp);
%
% ch1=jake(120,6);
%
%
% SP.paths(1,:)=tp(1)*ch1(1,:);
% SP.paths(2,:)=tp(2)*ch1(2,:);
% SP.paths(3,:)=tp(3)*ch1(3,:);
% SP.paths(4,:)=tp(4)*ch1(4,:);
% SP.paths(5,:)=tp(5)*ch1(5,:);
% SP.paths(6,:)=tp(6)*ch1(6,:);
%=====%======= Choose Equalization Type

% SP.equalizerType ='ZERO';
SP.equalizerType ='MMSE';


%%%%%%===================================

numSymbols = SP.FFTsize;
Q = numSymbols/SP.inputBlockSize;

    %%%%%%%%%========== Channel Generation============

      %%%%%%%%%==========Uniform channel============

%         SP.channel=SP.channel1(:,k).';

      %%%%%%%%%======= Vehicular A Channel ============
```

```
%        vechA_=[SP.paths(1,k) SP.paths(2,k) 0
SP.paths(3,k) SP.paths(4,k) 0 0 SP.paths(5,k) 0 0
SP.paths(6,k)];
%        SP.channel = vechA_;
%
        %%%%%%%%============ SUI3 channel============

    SUI3_1=[SP.paths(1,1) 0 0 SP.paths(2,1) 0
SP.paths(3,1)];
    SP.channel = SUI3_1/norm(SUI3_1);

%%%%%%%%================= AWGN channel============

%          SP.channel=1;

%%%%%%%%=========================================

        H_channel = fft(SP.channel,SP.FFTsize);

    % im=imread('lena.bmp');
im=imread('cameraman.tif');%image reading
%im=imread('D:\my work\Chaotic Mapping data\programs\
lena512.bmp');%im=imread('mri.tif');
xx=randomization(im);%image randomization

%*****************Data Generation********************
f=zeros(256,256);
f=xx;
[M,N]=size(f);
g=im2col(f, [M,N],'distinct');%image to column
converter
h=dec2bin(double(g));%pixel value to binary
conversion...every value replaced by 8 bits string
[M1,N1]=size(h) ;
z=zeros (M1,N1) ;
clear i j
for i=1:M1
for j=1:N1
z(i,j)= str2num(h(i,j)); %string to number conversion
end;
end;
[M2,N2] = size(z) ;
zz = reshape(z,M2*N2, 1);%parallel data reshaping date
to vector
```

```
% ********** Dividing the image into blocks***********

nloops = ceil((M2*N2)/SP.inputBlockSize );%number of
image %blocks by approximation
new_data = nloops*SP.inputBlockSize ;%new vector
proportional to %block size
nzeros   = new_data  - (M2*N2);%number of zeros to be
added to %old data vector
input_data = [zz;zeros(nzeros,1)];%construction of new
data %vector
input_data2 = reshape(input_data ,SP.
inputBlockSize ,nloops); %reshape the new data to matrix
of block size rows to number of %blocks columns
save input_data2

%************** transmission ON SC-FDMA **************

demodata1 = zeros(SP.inputBlockSize ,nloops);% this
matrix to %store received data vector

clear jj
for jj =1: nloops          % loop for columns
    b1= input_data2(:,jj)';%every block size

        %%%%%%%%%%%%%%% QPSK Modulation %%%%%%%%%%%%%%%%

          tmp = b1;
          tmp = tmp*2 - 1;
          inputSymbols = (tmp(1,:) + i*tmp(1,:))/
sqrt(2);

          %%%%%%%%%%%% SC-FDMA  Modulation %%%%%%%%%%%%

          inputSymbols_freq = fft(inputSymbols);
          inputSamples_ifdma = zeros(1,numSymbols);
          inputSamples_lfdma = zeros(1,numSymbols);

          %%%%%%%%%%%% Subcarriers Mapping %%%%%%%%%%%%

inputSamples_ifdma(1+SP.subband:Q:numSymbols) =
inputSymbols_freq;
      inputSamples_lfdma([1:SP.inputBlockSize]+SP.
inputBlockSize*SP.subband) = inputSymbols_freq;
      inputSamples_ifdma = ifft(inputSamples_ifdma);
      inputSamples_lfdma = ifft(inputSamples_lfdma);
```

```
%%%%%%%%%%%% Add Cyclic Prefix %%%%%%%%%%%%%

        TxSamples_ifdma = [inputSamples_
ifdma(numSymbols-SP.CPsize+1:numSymbols)
inputSamples_ifdma];
        TxSamples_lfdma = [inputSamples_
lfdma(numSymbols-SP.CPsize+1:numSymbols)
inputSamples_lfdma];

        %%%%%%%%%%%%% Wireless channel %%%%%%%%%%%%%%

        RxSamples_ifdma = filter(SP.channel, 1,
TxSamples_ifdma); % Multipath Channel
        RxSamples_lfdma = filter(SP.channel, 1,
TxSamples_lfdma); % Multipath Channel

        %%%%%%%%%%%%% Noise Generation %%%%%%%%%%%%%

        tmp = randn(2, numSymbols+SP.CPsize);
        complexNoise = (tmp(1,:) + i*tmp(2,:))/
sqrt(2);
        noisePower = 10^(-SP.SNR/10);

        %%%%%%%%%%%%%% Received signal%%%%%%%%%%%%%%

        RxSamples_ifdma = RxSamples_ifdma +
sqrt(noisePower/Q)*complexNoise;
        RxSamples_lfdma = RxSamples_lfdma +
sqrt(noisePower/Q)*complexNoise;

        %%%%%%%%%%%% Remove Cyclic Prefix%%%%%%%%%%%%%

        RxSamples_ifdma = RxSamples_ifdma(SP.
CPsize+1:numSymbols+SP.CPsize);
        RxSamples_lfdma = RxSamples_lfdma(SP.
CPsize+1:numSymbols+SP.CPsize);

        %%%%%%%%%%%%% SC-FDMA demodulation%%%%%%%%%%%

        Y_ifdma = fft(RxSamples_ifdma, SP.FFTsize);
        Y_lfdma = fft(RxSamples_lfdma, SP.FFTsize);

        %%%%%%%%%%% subcarriers demapping%%%%%%%%%%%%

        Y_ifdma = Y_ifdma(1+SP.subband:Q:numSymbols);
```

```
        Y_lfdma = Y_lfdma([1:SP.inputBlockSize]+SP.
inputBlockSize*SP.subband);

        %%%%%%%%%%%%%%%%% Equalization %%%%%%%%%%%%%%%%%%

        H_eff = H_channel(1+SP.subband:Q:numSymbols);
        if SP.equalizerType == 'ZERO'
            Y_ifdma = Y_ifdma./H_eff;
        elseif SP.equalizerType == 'MMSE'
            C = conj(H_eff)./(conj(H_eff).*H_eff +
10^(-SP.SNR/10));
            Y_ifdma = Y_ifdma.*C;
        end

        H_eff = H_channel([1:SP.inputBlockSize]+SP.
inputBlockSize*SP.subband);
        if SP.equalizerType == 'ZERO'
            Y_lfdma = Y_lfdma./H_eff;
        elseif SP.equalizerType == 'MMSE'
            C = conj(H_eff)./(conj(H_eff).*H_eff +
10^(-SP.SNR/10));
            Y_lfdma = Y_lfdma.*C;
        end

        EstSymbols_ifdma = ifft(Y_ifdma);
        EstSymbols_lfdma = ifft(Y_lfdma);

        %%%%%%%%%%%%%%%% demodulation%%%%%%%%%%%%%%%%%%%

        EstSymbols_ifdma = sign(real(EstSymbols_ifdma)) ;
        EstSymbols_ifdma =(EstSymbols_ifdma+1)/2;
        EstSymbols_lfdma = sign(real(EstSymbols_lfdma)) ;
        EstSymbols_lfdma = (EstSymbols_lfdma+1)/2;

        demodata1_ifdma(:,jj)   =   EstSymbols_
ifdma(:);   % the output of scfdma columns%storing of
received image data
        demodata1_lfdma(:,jj)  = EstSymbols_lfdma(:);
% the output of scfdma columns%storing of received
image data
end

 %****************  Received image  **************

 [M3,N3] = size(demodata1_ifdma);
```

```
%  demodata2 = demodata1(:);
 yy1_ifdma = reshape (demodata1_ifdma,M3,N3);
%reshaping the matrix to vector
 yy1_lfdma = reshape (demodata1_lfdma,M3,N3);
%reshaping the matrix to vector
 received_image_ifdma = yy1_ifdma(1:M2*N2);%taking the
original data
 received_image_lfdma = yy1_lfdma(1:M2*N2);%taking the
original data

%**************  Regeneration of image  **************

zz1_ifdma=reshape(received_image_ifdma,M2* N2,1);
%reshaping to M2*N2 vector
    zz1_lfdma=reshape(received_image_lfdma,M2* N2,1);
%reshaping to M2*N2 vector
yy_ifdma = reshape(zz1_ifdma,M2, N2);
yy_lfdma = reshape( zz1_lfdma,M2, N2);

clear i j
for i=1:M1
        for j=1:N1

zn_ifdma(i,j)=num2str(yy_ifdma(i,j));
zn_lfdma(i,j)=num2str(yy_lfdma(i,j));

end;
end;
hn_ifdma=bin2dec(zn_ifdma);
hn_lfdma=bin2dec(zn_lfdma);
gn_ifdma=col2im(hn_ifdma, [M,N], [M,N], 'distinct');
gn_lfdma=col2im(hn_lfdma, [M,N], [M,N], 'distinct');
y1_ifdma=derandomization(gn_ifdma);
 y1_lfdma=derandomization(gn_lfdma);
y1_ifdma=y1_ifdma/255;
 y1_lfdma= y1_lfdma/255;
% **************** The output results****************

figure (1)
imshow(im)
figure (2)
 imshow(y1_ifdma)
 MSE1_ifdma=sum(sum((double(im)/255-y1_ifdma).^2))/
prod(size(im));
PSNR_ifdma=10*log(1/MSE1_ifdma)/log(10);
```

```
 figure (3)
 imshow(y1_lfdma)
 MSE1_lfdma=sum(sum((double(im)/255-y1_lfdma).^2))/
prod(size(im));
PSNR_lfdma=10*log(1/MSE1_lfdma)/log(10);
toc
%PSNR(e)=10*log(1/MSE1)/log(10);% Peak signal-to-noise
ratio 10*log10((max possible pixel value of the
image)^2/MSE)
% end
% ebno=0:2:8;
% figure (3)
% plot(ebno,PSNR)


%***************** End of file *******************


%********************************************************
function x=randomization(f)
[M,N]=size(f) ;
g=im2col(f, [M,N], [M,N], 'distinct');
h=dec2bin(double(g));
[M1,N1]=size(h);
z=zeros (M1,N1) ;
for i=1:M1
for jj=1:N1
z(i,jj)=str2num(h(i,jj));
end;
end;
      h_chaot = zeros(size(z));

   for j=1:8
       HH=z(:,j);
       FF=reshape(HH,M,N);
       n = [10,5,12,5,10,8,14,10,5,12,5,10,8,14,10,5,12,
5,10,8,14,10,5,12,5,10,8,14];
       %n=[4,1,5,6];
         %n = [10,5,12,5,10,8,14,10,5,12,5,10,8,14];
%   n=[5 2 1 4 3 1];
%n=[4 2 4 4 2];
       [pr,pc] = chaomat(n);

       pim = chaoperm(FF,pr,pc,3,'forward');

       h_chaot(:,j)=reshape(pim,M*N,1);
   end
```

```
    for i=1:M1
for jj=1:N1
zn(i,jj)=num2str(h_chaot(i,jj));
end;
end;

hn=bin2dec(zn);
x=col2im(hn, [M,N], [M,N], 'distinct');

%****************************************************
function f=derandomization(x)
[M,N]=size(x) ;
g=im2col(x, [M,N], [M,N], 'distinct');
h=dec2bin(double(g));
[M1,N1]=size(h);
z=zeros (M1,N1) ;
for i=1:M1
for jj=1:N1
z(i,jj)=str2num(h(i,jj));
end;
end;
        h_rec = zeros(size(z));

    for j=1:8
        HH=z(:,j);
        FF=reshape(HH,M,N);
        n = [10,5,12,5,10,8,14,10,5,12,5,10,8,14,10,5,12,
5,10,8,14,10,5,12,5,10,8,14];
        %n=[4,1,5,6];
        %n = [10,5,12,5,10,8,14,10,5,12,5,10,8,14];
%   n=[5 2 1 4 3 1];
%n=[4 2 4 4 2];
        [pr,pc] = chaomat(n);
        pim = chaoperm(FF,pr,pc,3,'backward');
        h_rec(:,j)=reshape(pim,M*N,1);
    end

    for i=1:M1
for jj=1:N1
zn(i,jj)=num2str(h_rec(i,jj));
end;
end;

hn=bin2dec(zn);
f=col2im(hn, [M,N], [M,N], 'distinct');
```

```
%***********************************************************
function [pr,pc]=chaomat(n)
%
I=sum(n);
k=size(n,2);
for i=1:k
    N(i+1)=1;
    for j=1:i
        N(i+1)=N(i+1)+n(j);
    end
end
N(1)=1;
%N(0)=1;
for cb=1:k
    for rb=1:n(cb)
        rbstartcol(rb)=mod((rb-1)*I,n(cb));
        rbendcol(rb)=mod((rb*I-1),n(cb));
        rbstartrow(rb)=fix(((rb-1)*I)/n(cb));
        rbendrow(rb)=fix((rb*I-1)/n(cb));
        mincol(rb)=min([rbendcol(rb)+1,rbstartcol(rb)]);
        maxcol(rb)=max([rbendcol(rb),rbstartcol(rb)-1]);
    end

    for i=1:I
        for j=N(cb):N(cb+1)-1
            newindex(i,j-N(cb)+1)=(i-1)*n(cb)+
(n(cb)-j+N(cb)-1);
            newindexmod(i,j-N(cb)+1)=mod
(newindex(i,j-N(cb)+1),n(cb));
            newindexquotient(i,j-N(cb)+1)=fix(newindex
(i,j-N(cb)+1)/n(cb));
            rowblockindex(i,j-N(cb)+1)=fix(newindex
(i,j-N(cb)+1)/I)+1;
        end
    end

    for i=1:I
        for j=1:n(cb)
            for rb=1:n(cb)
                if rowblockindex(i,j)==rb;
                    if newindexmod(i,j)>maxcol(rb)
                        col=rbendrow(rb)-newindexquotient
(i,j)+(n(cb)-1-newindexmod(i,j))*(rbendrow(rb)-
rbstartrow(rb));
```

```
                    elseif newindexmod(i,j)>=mincol(rb) &
newindexmod(i,j)<=maxcol(rb)
                        if rbstartcol(rb)>rbendcol(rb)
                            c=0;
                            d=-1;
                        else
                            c=1;
                            d=1;
                        end
                        col=(rbendrow(rb)-
rbstartrow(rb))*(n(cb)-1-maxcol(rb))+(rbendrow(rb)-
newindexquotient(i,j)+c)+(maxcol
(rb)-newindexmod(i,j))*(rbendrow(rb)-rbstartrow(rb)+d);
                    else %if newindexmod(i,j)<=mincol(rb)
                        col=I-mincol(rb)*(rbendrow(rb)-
rbstartrow(rb))+(rbendrow(rb)-newindexquotient(i,j)+1)
+(mincol(rb)-1-newindexmod(i,j))*(rbendrow(rb)-
rbstartrow(rb));
                    end

                    row=1+I-N(cb+1)+rowblockindex(i,j);
                end
            end

            pr(i,j+N(cb)-1)=row;
            pc(i,j+N(cb)-1)=col;
        end
    end
end


%***************************************************
function out=chaoperm(im,pr,pc,num,forward)
%
[rows,cols] = size(im);
mat = zeros([rows,cols,num+1]);
mat(:,:,1) = im(:,:);

for loc=2:num+1
    if(strcmp(forward,'forward'))
        for i=1:rows
            for j=1:cols
                mat(pr(i,j),pc(i,j),loc) =
mat(i,j,loc-1);
            end
        end
```

```
      elseif(strcmp(forward,'backward'))
          for i=1:rows
              for j=1:cols
                  mat(i,j,loc) = mat(pr(i,j),
pc(i,j),loc-1);
              end
          end
      end
end
out = mat(:,:,num+1);
```

%**

```
function paths= fadchan(SP)
N=8
OR = 4;
M = 256;
Dop_res = 0.1;
res_accu = 20;

% P = [ 0 -5 -10 ];
% K = [ 0.1 0 0 ];
% tau = [ 0.0 5 10];
% Dop = [ 2 1.5 2.5];
% Dop = [ 0  0 0];

P = [ 0 -5 -10 ];
K = [ 1 0 0 ];
tau = [ 0.0 0.5 1.0 ];
Dop = [ 0.4 0.4 0.4 ];
ant_corr = 0.4;
Fnorm = -1.5113;
P = 10.^(P/10); % calculate linear power
s2 = P./(K+1); % calculate variance
m2 = P.*(K./(K+1)); % calculate constant power
m = sqrt(m2);
L = length(P); % number of taps
paths_r = sqrt(1/2)*(randn(L,N) + j*randn(L,N)).
*((sqrt(s2))' * ones(1,N));
paths_c = m' * ones(1,N);
for p = 1:L
D = Dop(p) / max(Dop) / 2; % normalize to highest
Doppler
f0 = [0:M*D]/(M*D); % frequency vector
PSD = 0.785*f0.^4 - 1.72*f0.^2 + 1.0; % PSD approximation
```

```
filt = [ PSD(1:end-1) zeros(1,M-2*M*D) PSD(end:-1:2)
]; % S(f)
filt = sqrt(filt); % from S(f) to |H(f)|
filt = ifftshift(ifft(filt)); % get impulse response
filt = real(filt); % want a real-valued filter
filt = filt / sqrt(sum(filt.^2)); % normalize filter
path = fftfilt(filt, [ paths_r(p,:) zeros(1,M) ]);
paths_r(p,:) = path(1+M/2:end-M/2);
end;
paths = paths_r + paths_c;
paths = paths * 10^(Fnorm/20); % multiply all
coefficients with F


%*********************************************************


function [Tt]=jake(v,N);
% v=120;
% N=6;
fc=4096000;
c=3*10^8;
v=v/3.6;
f=2*10^9; % carrier frequency
Tc=1/fc;%chip duration
nu=1;%1 samples per chip
fs=nu*fc; % sampling frequency
Ts=1/fs;
wM=2*pi*f*v/c;
%%%%%%%%%%%%%%%%%%
tet=rand(1,200)*2*pi;
t=0:128*Tc:5000*128*Tc;
No=16;
%%%%%%%%%%%%%%%%%%%%
A=hadamard(No);
%%%%%%%%%%%%%%%%%%%%
for k=1:N
Ttg=zeros(1,length(t));
tet=rand(1,200)*2*pi;
for l=1:No
B_n=pi*l/(No);
An=2*pi*(l-.5)/(4*No);
wn=wM*cos(An);
Ttg=A(k,l)*sqrt(2/No)*(cos(B_n)+j*sin(B_n)).*cos(wn*t+
tet(l))+Ttg;
end
Tt(k,:)=Ttg;
```

```
end
Tt=Tt(:,1:5000);

for z1=1:N
Tt(z1,:)=(Tt(z1,:)-mean(Tt(z1,:)))/std(Tt(z1,:));
end
%Tt=Tt(:,1:20);


%*******************************************************


%*******************************************************
% Simulation program to realize OFDM image transmission
%*******************************************************

%*****************preparation Part******************
clc
clear all

para=256;    % number of parallel channel to transmit
fftlen=256;  %FFT length
noc=256;     %number of carrier
nd=6;        %number of information OFDM symbol for one
loop
m1=2;        %Modulation level:QPSK
sr=250000;   %symbol rate
 br=sr.*m1;  %Bit rat per carrier
gilen=32;    %Length of guard interval (points)
ebno=6;      %Eb/No
Ipoint = 8; %Number of over samples
ofdm_length = para*nd*m1;    %Total no for one loop
modh=1/(2*pi);
A=1;


%************************Transmitter******************

%**********************Data Generation**************

%     im=imread('lena.bmp');
%im=imread('C:\Users\E. S. Hassan\Desktop\ima.gif');
%im=imread('C:\Documents and Settings\emad\
Desktop\29-10\programs\lena512.bmp');
im=imread('mri.tif');
f=double(im);
[M,N]=size(f);
```

```
g=im2col(f, [M,N], [M,N], 'distinct');
h=dec2bin(double(g));
[M1,N1]=size(h) ;
z=zeros (M1,N1) ;
for i=1:M1
for j=1:N1
z(i,j)= str2num(h(i,j));
end;
end;
[M2,N2] = size(z) ;
zz = reshape(z,M2*N2, 1);      %parallel data

% *********** Dividing the image into blocks**********
nloops = ceil((M2*N2)/ofdm_length );
new_data = nloops*ofdm_length ;
nzeros   = new_data  - (M2*N2);
input_data = [zz;zeros(nzercs,1)];
input_data2 = reshape(input_data ,ofdm_length ,
nloops);

%************** transmission ON OFDM ****************

demodata1 = zeros(ofdm_length ,nloops);
peakRMS1 = zeros(ofdm_length ,nloops);
peakRMS2 = zeros(ofdm_length ,nloops);

theta0=2*pi*rand(1,nloops)-pi; % memory terms (assume
uniform)
phi=zeros(ofdm_length,nloops); % initialize CE-OFDM
phase signal

for jj = 1: nloops            % loop for columns
    serdata1 = input_data2(:,jj)';
    [demodata,peakRMS1,peakRMS2]= ofdm_channelc(serdata1,
para,nd,m1,gilen,fftlen,sr,ebno, br);
%   demodata = ofdm_channel(serdata1,para,nd,m1,gilen,
fftlen,sr,ebno, br);
   phi(:,jj) = 2*pi*modh*demodata(:)+theta0(jj);   %
the output of ofdm columns
   peakRMS1(:,jj) = peakRMS1;
   peakRMS2(:,jj) = peakRMS2;
end

demodata1=A*exp(j*phi);
```

```
%***************** Received image **************
 [M3,N3] = size(demodata1);
% demodata2 = demodata1(:);
 yy1        = reshape (demodata1,M3*N3,1); %
 received_image   = yy1(1:M2*N2);


%************** Regeneration of image **************

yy = reshape(received_image  ,M2, N2);
zn=zeros (M1,N1) ;
for i=1:M1
for j=1:N1
zn(i,j)=num2str(yy(i,j));
end;
end;
hn=bin2dec(zn);
gn=col2im(hn, [M,N], [M,N], 'distinct');
gn=gn/255;
% ***************** The output results***************
 imwrite(gn,'image.tif', 'tif');
figure (1)
imshow(im)
figure (2)
 imshow(gn)
%************   The Error between Trans  ************

 MSE1=sum(sum((double(f)/255-gn).^2))/prod(size(f));
PSNR=10*log(1/MSE1)/log(10);


%*****************   End of file *******************


%*****************************************************


function [outdemodata,PRMS1,PRMS2 ]= ofdm_channelc
(serdata,para,nd,m1,gilen,fftlen,sr,ebno, br)

%   Simulation function to realize OFDM transmission
system
%
%*****************Input data*****************
% serdata : Input data as one vector
% para=128;   % number of parallel channel to
transmit
% fftlen=128; %FFT length
% noc=128;    %number of carrier
```

```
% nd=6;          %number of information OFDM symbol for
one loop
% m1=2;          %Modulation level:QPSK
% sr=250000;  %symbol rate
%   br=sr.*m1;  %Bit rat per carrier
% gilen=32;    %Length of guard interval (points)
%   ebno=3;       %Eb/No

%************serial to parallel conversion************
%   n = [10,5,12,5,10,8,14,10,5,12,5,10,8,14];
%   n=[5 2 1 4 3 1];
n=[4 2 4 4 2];
     [pr,pc] = chaomat(n);
paradata = reshape(serdata,para,nd*m1);

%*************** QPSK modulation ********************
[ich,qch] = qpskmod(paradata,para,nd,m1);

kmod = 1/sqrt(2);
ich1 = ich.*kmod;
qch1 = qch.*kmod;
%*************** IFFT ****************************

x = ich1 + qch1.*j;  % i :imag part
y = ifft(x);
ich2 = real (y);
qch2 = imag (y);
% before chaomat:.
%==================
%  for rr=1:length(icht)
Time_Signal  = sqrt(ich2.^2 + qch2.^2)./2;    % average
Time Signal
% peak_to_RMS(rr)  = 10^log10 (max(Time_Signal.^2)/
(std( Time_Signal )^2))
 peak_to_RMS  = 10^log10 (max(Time_Signal(:).^2)/
(std( Time_Signal(:))^2))

%===============generate the chao  ==================
[Rows,Cols] = size(ich2);
ichc = zeros (Rows,Cols);
qchc = zeros (Rows,Cols);
for jjj= 1:Cols
    im1= ich2(:,jjj);
    im1= col2im(im1, [16,16], [16,16], 'distinct'); %
im1 = reshape(im1,16,16);
```

```
      pim1 = chaoperm(im1,pr,pc,3,'forward');
      pim1 = im2col( pim1, [16,16], [16,16], 'distinct')
      ichc(:,jjj) = pim1;
      % the qch
      im2= qch2(:,jjj);
      im2= col2im(im2, [16,16], [16,16], 'distinct');
      pim2 = chaoperm(im2,pr,pc,3,'forward');
      pim2 = im2col( pim2, [16,16], [16,16], 'distinct');
      qchc(:,jjj) = pim2;
end
TimeSignal = sqrt(ichc.^2 + qchc.^2)/2;
% peaktoRMS (rr) = 10*log10(max(TimeSignal.^2)/
(std(TimeSignal)^2))
  peaktoRMS  = 10*log10(max(TimeSignal(:).^2)/
(std(TimeSignal(:))^2))

% calculate the PRMS:
 [L,Q] = size(Time_Signal);
PRMS1 = zeros (L,Q);
PRMS2 = zeros (L,Q);
for l= 1:L
    for q= 1:Q
 PRMS1 = 10^log10(max(Time_Signal(:).^2)/
(std( Time_Signal(:))^2));
 PRMS2 = 10*log10(max(TimeSignal(:).^2)/
(std(TimeSignal(:))^2));
    end
end

%***********  Guard interval insertion   ************

[ich3,qch3] = giins(ichc,qchc,fftlen,gilen,nd);
fftlen2 = fftlen + gilen;

% *********    Attenuation Calculation    **********

spow = sum(ich3.^2+qch3.^2)/nd./para;
attn = 0.5*spow*sr/br*10.^(-ebno/10);
attn = sqrt (attn);

%$$$$$$$$$$$$$$$$$$    Receiver       $$$$$$$$$$$$$$$$$$

%****************  AWGN addition   *****************

 [ich4,qch4] = comb(ich3,qch3,attn);
```

```
%************* Guard interval removal **************
[ich5,qch5] = girem (ich4,qch4,fftlen2,gilen,nd);
%[ich5,qch5] = girem (ich3,qch3,fftlen2,gilen,nd);
%=============% restore the original image===========

[Row2,Col2] = size(ich5)
ichr = zeros (Row2,Col2);
qchr = zeros (Row2,Col2);
for jjj  = 1:Col2
    im3  = ich5(:,jjj);
    im3  = col2im(im3, [16,16], [16,16], 'distinct');
    pim3 = chaoperm(im3,pr,pc,3,'backward');
    pim3 = im2col( pim3, [16,16], [16,16],
'distinct');
    ichr(:,jjj) = pim3(:);
    % the qch
    im4= qch5(:,jjj);
    im4= col2im(im4, [16,16], [16,16], 'distinct');
    pim4 = chaoperm(im4,pr,pc,3,'backward');
    pim4 = im2col( pim4, [16,16], [16,16],
'distinct');
    qchr(:,jjj) = pim4(:);
end

%********************* FFT  **********************

rx = ichr + (qchr.*j);
ry = fft(rx);
ich6 = real (ry);
qch6 = imag (ry);

%***************** Demodulation ******************
 ich7 = ich6./ kmod;
 qch7 = qch6./ kmod;
 outdemodata  = qpskdemod (ich7,qch7,para,nd,m1);
% output as one columns:
%==========================
%   outdemo  = qpskdemod (ich7,qch7,para,nd,m1);
%    outdemdata =  outdemo(:);
%  outdemdata = reshape (outdemo1,1,(para*nd*m1));

%********************** End **********************
```

References

1. R. V. Nee and R. Prasad, *OFDM for Wireless Multimedia Communications*, Artech House, Boston, MA, 2000.
2. R. Prasad, *OFDM for Wireless Communications Systems*, Artech House, Boston, MA, 2004.
3. H. Schulze and C. Luders, *Theory and Application of OFDM and CDMA: Wideband Wireless Communication*, Wiley, Chichester, U.K., 2005.
4. J. S. Chow, J. C. Tu, and J. M. Cioffi, A discrete multitone transceiver system for HDSL applications, *IEEE J. Sel. Areas Commun.*, 9, 895–908, August 1991.
5. H. Dai and H. V. Poor, Advanced signal processing for power line communications, *IEEE Commun. Mag.*, 41(5), 100–107, May 2003.
6. S. B. Weinstein, The history of orthogonal frequency division multiplexing, *IEEE Commun. Mag.*, 47, 26–35, November 2009.
7. E. S. Hassan, S. E. El-Khamy, M. I. Dessouky, S. A. El-Dolil, and F. E. Abd El-Samie, Peak to average power ratio reduction for OFDM signals with unequal power distribution strategy and the selective mapping technique, *Proceedings of NRSC-2010*, Menouf, Egypt, March 16–18, 2010.
8. A. Gusmão, R. Dinis, and N. Esteves, On frequency-domain equalization and diversity combining for broadband wireless communications, *IEEE Commun. Lett.*, 51(7), 1029–1033, July 2003.
9. D. Falconer, S. Ariyavisitakul, A. Benyamin-Seeyar, and B. Eidson, Frequency domain equalization for single-carrier broadband wireless systems, *IEEE Commun. Mag.*, 40(4), 58–66, April 2002.
10. F. Pancaldi, G. Vitetta, R. Kalbasi, N. Al-Dhahir, M. Uysal, and H. Mheidat, Single-carrier frequency domain equalization, *IEEE Signal Process. Mag.*, 25(5), 37–56, September 2008.

11. H. Sari, G. Karam, and I. Jeanclaude, Transmission techniques for digital terrestrial TV broadcasting, *IEEE Commun. Mag.*, 33, 100–109, February 1995.

12. X. Zhu and R. Murch, Layered space-frequency equalization in a single-carrier MIMO system for frequency-selective channels, *IEEE Trans. Wireless Commun.*, 3, 701–708, May 2004.

13. L. Hanzo, L. Yang, E. Kuan, and K. Yen, *Single and Multi-Carrier DS-CDMA: Multi-User Detection, Space-Time Spreading, Synchronisation, Networking and Standards*, Wiley, New York, 2003.

14. V. Aue and G. P. Fettweis, Multi-carrier spread spectrum modulation with reduced dynamic range, *Proceeding of the IEEE 46th Vehicular Technology Conference*, Atlanta, GA, April 28–May 1, 1996, vol. 2, pp. 914–917.

15. S. Hara and R. Prasad, Overview of multicarrier CDMA, *IEEE Commun. Mag.*, 35, 126–133, December 1997.

16. S. Verdu, *Multiuser Detection*, Cambridge University Press, Cambridge, U.K., 1998.

17. Y. Yoshida, K. Hayashi, H. Sakai, and W. Bocquet, Analysis and compensation of transmitter IQ imbalances in OFDMA and SC-FDMA systems, *IEEE Trans. Signal Process.*, 57(8), 3119–3129, August 2009.

18. E. S. Hassan, X. Zhu, S. E. El-Khamy, M. I. Dessouky, S. A. El-Dolil, and F. E. Abd El-Samie, Performance evaluation of OFDM and single-carrier systems using frequency domain equalization and phase modulation, *Int. J. Commun. Syst.*, 24, 1–13, 2011.

19. A. Elbehery, S. A. S. Abdelwahab, M. A. El Naby, E. S. Hassan, S. Elaraby, and F. E. Abd El-Samie, Efficient image transmission over the single carrier frequency division multiple access system using chaotic interleaving, *Proceedings of the 29th NRSC-2012*, Cairo, Egypt, April 10–12, 2012.

20. M. Sharif, M. Gharavi-Alkhansari, and B. H. Khalaj, On the peak-to-average power of OFDM signals based on oversampling, *IEEE Trans. Commun.*, 51, 72–78, January 2003.

21. G. Wunder and H. Boche, Upper bounds on the statistical distribution of the crest-factor in OFDM transmission, *IEEE Trans. Inform. Theory*, 49, 488–494, February 2003.

22. H. Myung, Single carrier orthogonal multiple access technique for broadband wireless communications, PhD dissertation, Polytechnic University, Brooklyn, NY, January 2007.

23. J. A. O'Sullivan, R. E. Blahut, and D. L. Snyder, Information-theoretic image formation, *IEEE Trans. Inform. Theory*, 44(6), 2094–2123, 1998.

24. S. Saraswathi and A. Allirani, Survey on image segmentation via clustering, *Proceedings of Information Communication and Embedded Systems (ICICES)*, S.A. Engineering College, Chennai, Tamilnadu, India, 2013, pp. 331–335. doi: 10.1109/ICICES.2013.6508376.

25. R. C. Gonzalez and R. E. Woods, *Digital Image Processing*, Prentice Hall, Upper Saddle River, NJ, 2002.

26. H. C. Andrews, Digital image processing, *Comput. J.*, 7(5), 36–45, 1974, doi: 10.1109/MC.1974.6323522.

27. R. Maani, S. Camorlinga, and N. Arnason, Transforming medical imaging applications into collaborative PACS-based telemedical systems, *SPIE MI*, 7967, 79670D–79670D-10, 2011.

28. G. Arslan, B. L. Evans, and S. Kiaei, Equalization for discrete multitone transceivers to maximize bit rate, *IEEE Trans. Signal Process.*, 49(12), 3123–3135, December 2001.

29. S. Celebi, Interblock interference (IBI) minimizing time-domain equalizer (TEQ) for OFDM, *IEEE Signal Process. Lett.*, 10(8), 232–234, August 2003.

30. Y. Wu and W. Y. Zou, Orthogonal frequency division multiplexing: A multi-carrier modulation scheme, *IEEE Trans. Consum. Electron.*, 41(3), 392–399, August 1995.

31. S. Hara and R. Prasad, *Multicarrier Techniques for 4G Mobile Communications*, Artech House, Boston, MA, 2003.

32. E. S. Hassan, X. Zhu, S. E. El-Khamy, M. I. Dessouky, S. A. El-Dolil, and F. E. Abd El-Samie, Enhanced performance of OFDM and single-carrier systems using frequency domain equalization and phase modulation, *Proceedings of the NRSC-09*, Cairo, Egypt, March 17–19, 2009.

33. S. C. Thompson and A. U. Ahmed, Constant-envelope OFDM, *IEEE Trans. Commun.*, 56, 1300–1312, August 2008.

34. M. V. Clark, Adaptive frequency-domain equalization and diversity combining for broadband wireless communications, *IEEE J. Sel. Areas Commun.*, 16(8), 1385–1395, October 1998.

35. A. Gusmão, P. Torres, R. Dinis, and N. Esteves, A reduced-CP approach to SC/FDE block transmission for broadband wireless communications, *IEEE Trans. Commun.*, 55(4), 801–809, 2007.

36. K. Fazel and S. Kaiser, *Multi-Carrier and Spread Spectrum Systems*, Wiley, Chichester, U.K., 2003.

37. A. S. Baiha, M. Singh, A. J. Goldsmith, and B. R. Saltzberg, A new approach for evaluating clipping distortion in multicarrier systems, *IEEE J. Sel. Areas Commun.*, 20(5), 1037–1046, June 2002.

38. R. V. Nee and A. D. Wild, Reducing the peak-to-average power ratio of OFDM, *Proceedings of the IEEE Vehicular Technology Conference (VTC'98)*, Ottawa, Ontario, Canada, May 1998, pp. 2072–2076.

39. S. Wei, D. L. Goeckel, and P. E. Kelly, A modern extreme value theory scheme to calculating the distribution of the PAPR in OFDM systems, *Proceedings of the IEEE ICC 2002*, New York, May 2002, pp. 1686–1690.

40. S. H. Han and J. H. Lee, Modified selected mapping scheme for PAPR reduction of coded OFDM signal, *IEEE Trans. Broadcast.*, 50, 335–341, September 2004.

41. A. A. M. Saleh, Frequency-independent and frequency-dependent nonlinear models of TWT amplifiers, *IEEE Trans. Commun.*, 29, 1715–1720, November 1981.

42. H. G. Ryu, T. P. Hoa, K. M. Lee, S. W. Kim, and J. S. Park, Improvement of power efficiency of HPA by the PAPR reduction and predistortion, *IEEE Trans. Consum. Electron.*, 50, 119–124, February 2004.

43. T. Jiang and Y. Wu, An overview: Peak-to-average power ratio reduction techniques for OFDM signals, *IEEE Trans. Broadcast.*, 54(2), 275–268, June 2008.

44. X. Li, and L. J. Cimini, Effects of clipping and filtering on the performance of OFDM, *IEEE Commun. Lett.*, 2(5), 131–133, May 1998.

45. S. Deng and M. Lin, OFDM PAPR reduction using clipping with distortion control, *IEEE Int. Conf. Commun.*, 4, 2563–2567, May 2005.

46. S. Cha, M. Park, S. Lee, K. Bang, and D. Hong, A new PAPR reduction technique for OFDM systems using advanced peak windowing method, *IEEE Trans. Consum. Electron.*, 54(2), 405–410, May 2008.

47. Y. Lee, Y. You, W. Jeon, J. Paik, and H. Song, Peak-to-average power ratio in MIMO-OFDM systems using selective mapping, *IEEE Commun. Lett.*, 7, 575–577, December 2003.

48. M. S. Beak, M. J. Kim, Y. H. You, and H. K. Song, Semi-blind estimation and PAR reduction for MIMO-OFDM system with multiple antennas, *IEEE Trans. Broadcast.*, 50, 414–424, December 2004.

49. H. Chen and H. Liang, Combined selective mapping and binary cyclic codes for PAPR reduction in OFDM systems, *IEEE Trans. Wireless Commun.*, 6, 3524–3528, October 2007.

50. A. Ghassemi and T. A. Gulliver, Partial selective mapping OFDM with low complexity IFFTs, *IEEE Commun. Lett.*, 12, 4–6, January 2008.

51. L. J. Cimini, Jr. and N. R. Sollenberge, Peak-to-average power ratio reduction of an OFDM signal using partial transmit sequences, *IEEE Commun. Lett.*, 4, 86–88, March 2000.

52. E. S. Hassan, S. E. El-Khamy, M. I. Dessouky, S. A. El-Dolil, and F. E. Abd El-Samie, Peak-to-average power ratio reduction in space–time block coded multi-input multi-output orthogonal frequency division multiplexing systems using a small overhead selective mapping scheme, *IET Commun.*, 3(10), 1667–1674, 2009.

53. S. C. Thompson, J. G. Proakis, and J. R. Zeidler, Noncoherent reception of constant envelope OFDM in flat fading channels, *Proceeding of the IEEE 16th International Symposium on Personal, Indoor and Mobile Radio Communication*, Berlin, Germany, 2005, pp. 517–521.

54. M. Jankiraman, *Space-Time Codes and MIMO Systems*, Artech House, Boston, MA, 2004.

55. B. Vucetic and J. Yuan, *Space-Time Coding*, Wiley, West Sussex, U.K., 2003.

56. D. Gesbert, M. Shafi, D. Shiu, P. J. Smith, and A. Naguib, From theory to practice: an overview of MIMO space-time coded wireless systems, *IEEE J. Sel. Areas Commun.*, 21(3), 281–301, April 2003.

57. L. Zheng and D. N. Tse, Diversity and multiplexing: A fundamental tradeoff in multiple-antenna channels, *IEEE Trans. Inform. Theory*, 49(5), 1073–1096, May 2003.

58. V. Tarokh, H. Jafarkhani, and A. R. Calderbank, Space-time block codes from orthogonal designs, *IEEE Trans. Inform. Theory*, 45(5), 1456–1467, July 1999.

59. S. M. Alamouti, A simple transmit diversity technique for wireless communications, *IEEE J. Sel. Areas Commun.*, 16(8), 1451–1458, October 1998.

60. G. J. Foschini, Layered space-time architecture for wireless communication in a fading environment when using multi-element antennas, *Bell Labs Tech. J.*, 1(2), 41–59, 1996.

61. V. Tarokh, H. Jafarkhani, and A. R. Calderbank, Space-time block coding for wireless communications: Performance results, *IEEE J. Sel. Areas Commun.*, 17(3), 451–460, March 1999.

62. H. Yang, A road to future broadband wireless access: MIMO-OFDM-based air interface, *IEEE Commun. Mag.*, 43(1), 53–60, January 2005.

63. S. H. Han and J. H. Lee, A new PTS OFDM scheme with low complexity for PAPR reduction, *IEEE Trans. Broadcast.*, 52, 77–82, 2006.

64. S. B. Slimane, Reducing the peak-to-average power ratio of OFDM signals through precoding, *IEEE Trans. Veh. Techn.*, 56, 686–695, March 2007.

65. R. W. Baml, R. F. Fisher, and J. B Huber, Reducing the peak-to-average power ratio of multicarrier modulation by selected mapping, *IEE Electron. Lett.*, 32, 2056–2057, 1996.

66. A. D. Jayalath and C. Tellambura, SLM and PTS peak-power reduction of OFDM signals without side information, *IEEE Trans. Wireless Commun.*, 4, 2006–2013, 2005.

67. E. S. Hassan, S. E. El-Khamy, M. I. Dessouky, S. A. El-Dolil, and F. E. Abd El-Samie, A simple selective mapping algorithm for the peak to average power ratio in space time block coded MIMO-OFDM systems, *Proceedings of the HPCNCS-08*, Orlando, FL, July 7–10, 2008.

68. S. Han and J. Lee, An overview of peak-to-average power ratio reduction techniques for multicarrier transmission, *IEEE Trans. Wireless Commun.*, 12, 56–65, April 2005.

69. R. F. H. Fischer and M. Hoch, Directed selected mapping for peak-to-average power ratio reduction in MIMO OFDM, *IET Electron. Lett.*, 42(22), 1289–1290, October 2006.

70. R. F. H. Fischer and M. Hoch, Peak-to-average power ratio reduction in MIMO OFDM, *Proceedings of the IEEE ICC*, Scottish Exhibition & Conference Centre (SECC), Glasgow, Scotland, June 2007, pp. 762–767.

71. E. Alsusa and L. Yang, Redundancy-free and BER-maintained selective mapping with partial phase-randomising sequences for peak-to-average power ratio reduction in OFDM systems, *IET Commun.*, 2, 66–74, 2008.

72. T. Jiang, M. Guizani, H. Chen, W. Xiang, and Y. Wu, Derivation of PAPR distribution of the peak-to-average power ratio in OFDM signals, *IEEE Trans. Wireless Commun.*, 7, 1298–1305, April 2008.

73. C. Wang and Y. Ouyang, Low-complexity selected mapping schemes for peak-to-average power ratio reduction in OFDM systems, *IEEE Trans. Signal Process.*, 53(12), 4652–4660, December 2005.

74. E. S. Hassan, S. E. El-Khamy, M. I. Dessouky, S. A. El-Dolil, and F. E. Abd El-Samie, PAPR reduction for OFDM signals with unequal power distribution strategy and a reduced-complexity SLM scheme, *J. Central South Univ. Technol., Springer*, 19(7), 1902–1908, July 2012.

75. L. Yang, K. Soo, Y. M. Siu, and S. Q. Li, A low complexity selected mapping scheme by use of time domain sequence superposition technique for PAPR reduction in OFDM system, *IEEE Trans. Broadcast.*, 54(4), 821–824, December 2008.

76. E. Alsusa and L. Yang, Selective post-IFFT amplitude randomising for peak-to-average power ratio reduction in orthogonal frequency-division multiplexing based systems, *IET Commun.*, 2(4), 553–561, 2008.

77. F. Kohandani and A. Khandani, A new algorithm for peak/average power reduction in OFDM systems, *IEEE Trans. Brodcast.*, 54, 159–165, March 2008.

78. Y. Jie, C. Lei, L. Quan, and C. De, A modified selected mapping technique to reduce the peak-to-average power ratio of OFDM signal, *IEEE Trans. Consum. Electron.*, 53(3), 846–851, August 2007.

79. S. Wang and C. Li, A low-complexity PAPR reduction scheme for SFBC MIMO-OFDM systems, *IEEE Signal Proc.*, 16(11), 941–944, November 2009.

80. E. S. Hassan, X. Zhu, S. E. El-Khamy, M. I. Dessouky, S. A. El-Dolil, and F. E. Abd El-Samie, A continuous phase modulation single-carrier wireless system with frequency domain equalization, *Proceedings of the ICCES-09*, Cairo, Egypt, December 14–16, 2009.

81. S. C. Thompson, A. U. Ahmed, J. G. Proakis, and J. R. Zeidler, Constant envelope OFDM phase modulation: Spectral containment, signal space properties and performance, *Proceedings of the IEEE Milcom*, Monterey, CA, October 2004, vol. 2, pp. 1129–1135.

82. M. Kiviranta, A. Mammela, D. Cabric, D. A. Sobel, and R. W. Brodersen, Constant envelope multicarrier modulation: Performance evaluation in AWGN and fading channels, *Proceedings of the IEEE Milcom*, Atlantic City, NJ, October 2005, vol. 2, pp. 807–813.

83. Y. Tsai, G. Zhang, and J.-L. Pan, Orthogonal frequency division multiplexing with phase modulation and constant envelope design, *Proceedings of the IEEE Milcom*, Atlantic City, NJ, October 2005, vol. 4, pp. 2658–2664.

84. J. G. Proakis and D. G. Manolakis, *Digital Signal Processing: Principles, Algorithms, and Applications*, 3rd edn., Prentice Hall, Upper Saddle River, NJ, 1996.

85. J. G. Proakis and M. Salehi, *Communication Systems Engineering*, Prentice Hall, Upper Saddle River, NJ, 1994.

86. Y. Liu and Y. Du, A new coding scheme in SC-FDE, *Proceedings of the IEEE Wicom*, Wuhan, China, September 2006, pp. 1–4.

87. J. Li, Y. Du, and Y. Liu, Comparison of spectral efficiency for OFDM and SC-FDE under IEEE 802.16 scenario, *Proceedings of the IEEE ISCC'06*, Sardinia, Italy, June 2006, pp. 467–471.

88. Y. Wang, X. Dong, P. Wittke, and M. Shaomin, Cyclic prefixed single carrier transmission in ultra-wideband communications, *IEEE Trans. Wireless Commun.*, 5, 2017–2021, August 2006.

89. A. Barbieri, D. Fertonani, and G. Colavolpe, Spectrally efficient continuous phase modulations, *IEEE Trans. Wireless Commun.*, 8(3), 535–540, 2009.

90. D. J. Castello, J. Hagenauer, H. Imai, and S. Wicker, Applications of error-control coding, *IEEE Trans. Inform. Theory*, 44, 2384–2415, October 1998.

91. Y. Q. Shi, X. M. Zhang, Z.-C. Ni, and N. Ansari, Interleaving for combating bursts of errors, *IEEE Circuits Syst. Mag.*, 4 (First Quarter), 29–42, 2004.

92. V. D. Nguyen and H. Kuchenbecker, Block interleaving for soft decision Viterbi decoding in OFDM systems, *IEEE VTC*, Vol. 1, 470–474, 2001.

93. B. Jovic and C. Unsworth, Chaos-based multi-user time division multi-plexing communication system, *IET Commun.*, 1(4), 549–555, 2007.

94. R. Matthews, On the derivation of a chaotic encryption algorithm, *Cryptologia XIII*, 1, 29–42, 1989.

95. K. S. Deffeyes, Encryption system and method, US Patent No. 5001754, March 1991.

96. J. Fridrich, Symmetric ciphers based on two-dimensional chaotic maps, *Int. J. Bifurcat. Chaos*, 8, 1259–1284, October 1998.

97. F. Han, X. Yu, and S. Han, Improved baker map for image encryption, *ISSCAA*, Harbin, China, 2006, pp. 1273–1276.

98. E. S. Hassan, S. E. El-Khamy, M. I. Dessouky, S. A. El-Dolil, and F. E. Abd El-Samie, New interleaving scheme for continuous phase modulation based OFDM systems using chaotic maps, *Proceedings of the WOCN-09*, Cairo, Egypt, April 28–30, 2009.

99. E. S. Hassan, S. E. El-Khamy, M. I. Dessouky, S. A. El-Dolil, and F. E. Abd El-Samie, A chaotic interleaving scheme for continuous phase modulation based single-carrier frequency-domain equalization systems, *Wireless Personal Commun.*, 62(1), 183–199, January 2012.

100. E. S. Hassan, S. E. El-Khamy, M. I. Dessouky, S. A. El-Dolil, and F. E. Abd El-Samie, A chaotic interleaving scheme for continuous phase modulation based OFDM systems, *Int. J. Electron.*, 100(1), 48–61, January 2013.

101. E. S. Hassan et al., Chaotic interleaving scheme for single-and multi-carrier modulation techniques implementing continuous phase modula-tion, *J. Franklin Institute*, 350(4), 770–789, May 2013.

102. T. S. Rappaport, *Wireless Communications Principles and Practice*, 2nd edn., Pearson Education, Upper Saddle River, NJ, 2002.

103. P. P. Dang and P. M. Chau, Robust image transmission over CDMA channels, *IEEE Trans. Consum. Electron.*, 46(3), 664–672, August 2000.

104. T. Kathiyaiah and T. H. Oh, Performance analysis of JPEG2000 trans-mission through low SNR MC-CDMA channel, *Proceedings of IEEE Ninth International Conference on Communications*, Kuala Lumpur, Malaysia, December 15–17, 2009.

105. J. K. Rogers and P. C. Cosman, Wavelet Zerotree image compression with packetization, *IEEE Signal Process. Lett.*, 5(5), 105–107, May 1998.

106. J. G. Proakis, *Digital Communications*, 3rd edn., McGraw-Hill, New York, 1995.

107. E. M. El-Bakary, E. S. Hassan, O. Zahran, S. A. El-Dolil, and F. E. Abd El-Samie, Efficient image transmission with multi-carrier CDMA, *Wireless Personal Commun.*, 69(2), 979–994, 2013.

108. A. Klein, Data detection algorithms specially designed for the downlink of CDMA mobile radio systems, *Proc. IEEE VTC*, 1, 203–207, 1997.

109. J. A. Davis and J. Jedwab, Peak-to-mean power control in OFDM, Golay complementary sequences, *IEEE Trans. Inform. Theory*, 45, 2397–2417, November 1999.

110. J. Armstrong, Peak-to-average power reduction for OFDM by repeated clipping and frequency domain filtering, *IEE Electron. Lett.*, 38, 246–247, February 2002.

111. H. G. Myung, J. Lim, and D. Goodman, Peak-to-average power ratio of single carrier FDMA signals with pulse shaping, *Proceedings of the IEEE International Symposium on Personal, Indoor and Mobile Radio Communications (PIMRC)*, Helsinki, Finland, September 2006.

112. U. Sorger, I. D. Broeck, and M. Schnell, Interleaved FDMA—A new spread spectrum multiple-access scheme, *IEEE International Conference on Communications (ICC'98)*, Atlanta, GA, Vol. 2, 1998, pp. 1013–1017.

113. H. Myung, J. Lim, and D. Goodman, Single carrier FDMA for uplink wireless transmission, *IEEE Veh. Tech. Mag.*, 1, 30–38, September 2006.

114. M. Wylie-Green, E. Perrins, and T. Svensson, Design and performance of a multiple access CPM-SC-FDMA transmission scheme, *IEEE International Waveform Diversity and Design (WD & D) Conference*, Radisson Resort Worldgate, Kissimmee, FL, 2009, pp. 286–290.

115. G. D. Mandyam, Sinusoidal transforms in OFDMA system, *IEEE Trans. Broadcast.*, 50(2), 172–184, June 2004.

116. P. Tan and N. C. Beaulieu, A comparison of DCT-based OFDM and DFT-based OFDM in frequency offset and fading channels, *IEEE Trans. Commun.*, 54(11), 2113–2125, November 2006.

117. F. S. Al-Kamali, M. I. Dessouky, B. M. Sallam, F. Shawki, and F. E. Abd El-Samie, A new single carrier FDMA system based on the discrete cosine transform, *Proceedings of the ICCES'9 Conference*, Cairo, Egypt, December 14–16, 2009, pp. 555–560.

118. A. Elbehery, S. A. S. Abdelwahab, M. A. El Naby, E. S. Hassan, S. Elaraby, and F. E. Abd El-Samie, Image transmission with DCT based SC-FDMA system using continuous phase modulation, *Proceedings of the NRSC*, Egypt, 2013.

119. F. E. Abd El-Samie, F. S. Al-Kamali, A. Y. Al-Nahari, and M. I. Dessouky, *SC-FDMA for Mobile Communications*, CRC Press, Boca, Raton, FL, 2013.

120. H. G. Myung and D. J. Goodman, *Single Carrier FDMA: A New Air Interface for Long Term Evolution*, Wiley, Chichester, U.K., 2008.

Index